Films of Fact

Nonfictions is dedicated to expanding and deepening
the range of contemporary documentary studies.
It aims to engage in the theoretical conversation
about documentaries, open new areas of scholarship,
and recover lost or marginalised histories.

General Editor, Professor Brian Winston

Other titles in the *Nonfictions* series:

Direct Cinema: Observational Documentary and the Politics of the Sixties
by Dave Saunders

Projecting Migration: Transcultural Documentary Practice
edited by Alan Grossman and Àine O'Brien

The Image and the Witness: Trauma, Memory and Visual Culture
Edited by Frances Guerin and Roger Hallas

Vision On: Film, Television and the Arts in Britain
by John Wyver

Building Bridges: The Cinema of Jean Rouch
edited by Joram ten Brink

Forthcoming titles in the *Nonfictions* series:

Documentary Display: Re-Viewing Nonfiction Film and Video
by Keith Beattie

Chávez: The Revolution Will Not Be Televised
– A Case Study in Politics and the Media
by Rod Stoneman

Written to mark the centenary of Paul Rotha, Documentarist
1907–1984

Films of Fact

A HISTORY OF SCIENCE IN DOCUMENTARY FILMS AND TELEVISION

Timothy Boon

WALLFLOWER PRESS LONDON & NEW YORK

sciencemuseum wellcometrust

First published in Great Britain in 2008 by
Wallflower Press
6 Market Place, London W1W 8AF
www.wallflowerpress.co.uk

A catalogue record for this publication is available from the British Library.

ISBN 978-1-905674-37-4 (pbk)
ISBN 978-1-905674-38-1 (hbk)

Book design by Elsa Mathern

Printed and bound in Poland; produced by Polskabook

CONTENTS

ACKNOWLEDGEMENTS

I owe a great debt to my employers at the Science Museum for permitting me leave of absence to complete the research and to write this book. I should like to thank in particular Jon Tucker and Heather Mayfield who, beyond normal limits, made it possible for me to be absent and to Robert Bud, Andrew Nahum and Peter Morris, who have been truly collegial fellow travellers over many years. This work would have been impossible without the generosity of the Wellcome Trust who awarded me the Research Leave Grant (078700/Z/05/Z) that gave me the liberty to pursue the extra lines of enquiry that have turned this work from isolated studies into the developed narrative it now represents. Ian Christie kindly arranged for me to work in the School of History of Art, Film and Visual Media at Birkbeck College, a most congenial setting. Nick Lambert has not only been a good colleague but, from his mine of expertise, has also helped with the production of the illustrations for the book. Wallflower Press have been a delight to work with, and Damon McCollin-Moore, from the Science Museum's Trading Company, has been patient and supportive.

The final version of the text has benefited from close reading by several generous colleagues and friends: Robert Bud, Ian Christie, John Christie, Ralph Desmarais, Allan Jones, Luke McKernan, Andrew Nahum, Patrick Russell and David Smith all read parts of the text and the book is significantly stronger for their perceptive comments. Martin Gorst read every chapter in the order I completed them and his sense of the demands of the narrative has saved the reader several blind turnings along the way. Brian Winston's close and generous reading has ensured the focus and consistency of the narrative. Tom Boon kindly read the proofs and assisted the clarity of the text considerably. Jane Davies assisted with the index. Any errors of fact or interpretation in the final text are, of course, mine.

Staff at the National Film and Television Archive and British Film Institute Library in London have supported this project, not just in the intensive year of its writing but for many years before, by making available for viewing the films that feature in these pages. In this connection I thank Kathleen Dickson, Steve Tollervy, Steve Foxon, Bryony Dixon, Janet Moat and Sonia Mullet, but particularly Patrick Russell who has been a generous and insightful colleague, as we developed the film programme that marked not just Paul Rotha's centenary, but that of four other documentarists too.

The book would have been impossible without the help of librarians at the Science Museum, Wellcome Trust and British Library, and archivists at the National Archives, Rotha Archive, Reading University Library, London School of Economics Library, RIBA Archive, UEA Archive, Grierson Archive, National Library of Scotland and Huxley Archive; permission to cite unpublished materials from collections in their care is gratefully acknowledged, along with that from the Galton Institute and the Policy Studies Institute for collections held at the Wellcome Library and the London School of Economics respectively. Particular thanks go to Erin O'Neill, archivist at the BBC Written Archives Centre, who responded cheerfully to an extended series of abstruse requests for files on long forgotten BBC programmes. Martin Freeth, David Dugan, Adam Curtis and Toni Charlton generously helped me to a better understanding of parts of the story beyond the archive. For permission to quote unpublished materials I thank in particular Robert Kruger, administrator of the Rotha estate. Several of my previous papers served as starting points for particular chapters: chapter two derives in part from my 2000 paper; chapter three owes much to that published in 1993, whilst the *World of Plenty* section of chapter four was first explored in 1997. I am grateful to the editors and publishers of these volumes (respectively, the British Society for the History of Science, Chris Lawrence and Anna Mayer, David Smith) for permission to re-publish some of this material, thoroughly re-written here.

Permission to reproduce pictures from the following sources is gratefully acknowledged. Plates 1.1–1.3: Science Museum/Science & Society Picture Library. Plates 2.1, 7.3: Museum of Science & Industry, Manchester. Plates 2.2–2.6: Viewtech Educational Media. Plate 2.7: Royal Mail Educational Services. Plate 2.8: Shell Film Unit. Plates 2.9, 3.1–3.5, 4.6: National Grid Archive. Plates 4.1–4.5, 5.1–5.7: Crown Copyright. Plates 6.1–6.2, 7.1, 7.4: BBC Information and Archives. Plate 8.1: Windfall Films. Pictures supplied by copyright holders or, with permission, from DVD copies, with the exception of: 2.7–9, 4.2–4, 5.7, supplied by BFI Stills Library.

Words are not enough to express my gratitude to Edith, Wilfrid, Nathaniel, and above all Alison, who have not visibly minded my absorption in things long ago and – until now – long forgotten.

PREFACE

Relations between science and the public, for those that follow the debate, are today at a critical point. Majority opinion amongst science communicators is in agreement that for science to retain democratic support, rectifying a deficit in public scientific knowledge will not suffice: engagement and dialogue between science and non-scientists is seen to be crucial. At the same time, the science in science filmmaking, if you agree with erstwhile *Horizon* editor Peter Goodchild (2004), is being sidelined for human drama, sensation or simply ratings. Even worse, non-fiction filmmaking is said to be in crisis, attacked at its realist roots by the fungus of postmodernism.

Could these crises in science engagement and in non-fiction filmmaking have a common cause? At the source of all of them are questions of authority. In the first case, society will no longer permit simple deference to the authority of scientists: the issues are too important. In the second, the authority is inverted and the public – filtered via world media markets – are calling the tunes. In the third, the authority of factual representation as a route to truth is in question (see Winston 1995). All three, recognisably products of the social and cultural revolutions of the last half century, are fine meat for critical analysis by the academy. But, this book proposes, we stand no chance of understanding the current conundrum in science television if we do not understand how we came to have the genres that characterise television science. That requires us to understand how they came out of their own specific historical circumstances, which is the subject of this book. Science and filmmaking are old – not odd – bedfellows, whether we take that assertion at the level of the epistemic guarantee of documentary realism, or in this book's sense of the old alliance between scientists and filmmakers in promoting – they might say 'popularising' – science. An understanding of the history of their interaction is an essential prerequisite in coming to a conclusion about whether the current state of science television is genuinely a matter for concern.

At the beginning of our story, it is clear that cinema, in its origins, owed much to particular nineteenth-century modes of science and innovation. We shall see by its conclusion that twentieth-century science decided long ago that it needs films, or at least television. These were films made for their particular present, not for us or for posterity. They may now be judged to be more or less powerful examples of film art, but their real significance lies in the special level of access to the past they provide, to previous attempts to persuade the public of the significance of science, technology and medicine.

In memory of my father, George C. Boon, and for my mother, Diana Boon.
A curious manifestation, perhaps, of filial affection.

INTRODUCTION

The title of ths book, *Films of Fact: A History of Science in Documentary Films and Television,* was chosen carefully. 'Films of fact' denotes territory that is not captured by the terms often used as synonyms: 'documentary', 'non-fiction' or 'educational' films. 'Documentary' has specific historical meanings that are analysed in chapter two, but it is only one factual genre of several. 'Non-fiction' is ambiguous, first in the pedantic sense that many factual films use fictional scenarios and, more philosophically, because nothing absolute separates pure fiction from the constructed narratives of factual films – there is a continuum. This book is, avowedly, concerned with science films aimed at the adult public, not with school films, so to use the term 'educational' would be plainly misleading. But 'Films of Fact', in addition to denoting the general field, was the name of one of Paul Rotha's film companies, and Rotha is a very significant individual for this account.

The history of the relationship between non-fiction films or television and science has until now received no extended treatment. This is the first book to build on focused case studies to present a way of thinking about scientific films made for the public that could be extended to any of the thousands of examples that I do not discuss here. All the same, this book is limited to Britain; we may look forward to synthetic accounts on a global scale, but the secondary literature is not yet strong enough to enable this.

In the view of Jack Ellis, John Grierson's most recent biographer, accounts of British documentary tend to fall into two camps: those that broadly align themselves with Grierson and the aims of documentary, and those that are seen to be critical of it, notably those that take issue with documentarists' various beliefs about realistic representation (for example, Ellis 2000: 332–3). In fact, this characterisation misses the point that many of the latter accounts belong not to history, but to different critical and philosophical traditions that are often greater in their theoretical ambitions as they deconstruct the truth claims of the realist filmmakers (see Winston 1995) or propose ideal-typical classifications for different kinds of documentary (see Nichols 1991; Plantinga 1997). Some of this work is also of practical value in addressing how better films might be made. This account, by contrast, is aimed to assist understanding of the past, to show what filmmakers and scientists were doing when they made films. It starts from the assumption that the historian should be neither aligned nor counter-aligned with their subject. In this sense it is a 'history' in the older sense of a being a naturalistic account of a category of human activity.

Films of Fact is compatible with some of the best existing work on science television. The potent contributions of Roger Silverstone (1985, 1986, 1987), Robert Young (1986; Gardner & Young 1981) and Harry Collins (1987) illuminate and critique the state of science television in the 1980s and beyond. This book, by contrast, shows the genealogy of the major genres of non-fiction scientific films and television. Or, to put it another way, Silverstone's *Framing Science* looks back to Aubrey Singer's statement, made in 1966, of the relationship between science and television, whereas this account shows how this was the *conclusion* of long historical processes (Silverstone 1985: 160-80).

Today, the cultural importance of science is everywhere asserted, but the cultural history of science is a comparatively new field. Many more studies of how culture can be found within science exist than the reverse, works exploring the presence and operation of science within culture. And it is a matter of democratic concern that the institutions that shape and determine how science and culture are represented are generally poorly understood. What Robert Young, speaking of the series *Crucible* (1982), said of science on television is true of this whole subject:

> The future – including the very existence – of civilization depends on getting right the relations between expertise and democracy. Television is the most pervasive and powerful medium for presenting issues in the public domain. Put these two assertions together and the attraction of a TV series on the social relations of science, technology and medicine – broadly conceived – is irresistible. (Young 1986: 3)

Wherever science is put together with powerful media of public communication, there is a potent conjunction, a nexus for the play of social power. The many ways that science may be represented – as the film itself opening the viewer's eyes to nature, as the source of transformative technologies large and small, as the route to human equality, or as a philosophical approach to the world – are contingent on their conditions of production and significant for the intellectual wellbeing of their societies.

'Science' itself is an ambiguous term, but here – at least in close description – we are assisted by reproducing the words that my story's participants used (which historians of science call 'actors' categories'). But we may note that they very often combined technology, science, medicine and rationality more generally under the heading 'science'.

The fate of genres in scientific filmmaking

Films of Fact takes specific examples of films and shows how they have arisen and been constructed within particular historical conditions. Each such product of local circumstances, I suggest, may seem peculiarly well suited – in the moment – to the exposition of a particular subject. If it is copied, or the techniques are applied on further occasions, we may, with care, speak of such a group as a 'genre' of factual film, as a term denoting the composite of techniques and representational conventions that several examples have in common.[1] Accordingly, each of the chapters that follow contains at least one example analysed in significant detail to demonstrate how it was the product of its circumstances. These examples are compared with other films on related themes from the specific period. On a larger scale, the book as a whole looks at the careers of such 'genres', showing how some persisted as their particular combinations of representational technique and approach to science, technology or medicine continued to seem relevant to filmmakers, funders and others in the extended networks responsible for scientific films. Other genres were replaced by the new forms produced in new circumstances. I do not argue that it is necessary for succeeding filmmakers to be conscious in copying previous genres, although that may sometimes be the case. Rather, filmmaking has most often been more akin to a craft practice in which at different times a particular approach has seemed to be appropriate or even correct; the source of that assumption has often not been clear.

Hayden White has suggested that historians tend to prefigure their accounts – to imagine their shape – according to widely available forms of emplotment: tragedy, comedy, romance or satire (White 1974: 7). The 'genres' I describe (particular combinations of aural, visual and structural tropes combined with particular emphases within the science represented) have been available, like White's forms of emplotment, to prefigure the accounts of science, technology and medicine in later films and programmes. They become part of the, usually unconscious, 'common-sense' of film and programme makers, affecting the selection of what seems to make sense within an account, aiding the selection of particular contents and the elision of other material. It is not, however, necessary to see these genres as anything other than the contingent products of historical circumstances, although at least one of them – Rotha's use of dialectical montage – had a genealogy that can be traced to the eighteenth century.

Scientific films made for the public have never been a genre entirely separate from other types of filmmaking. Of course, there has been a subculture of specifically scientific filmmaking; producers and directors have often specialised in making films about science, and there have been organisations –

such as the Scientific Film Association (see chapter four) – devoted to their encouragement. But, all the same, the first science films were types of kinematographic novelty act, scientific documentaries are types of documentary and television science programmes are a sub-genre of non-fiction programming. By contrast, *research* films, using the cine-camera as a scientific instrument, are a separate and distinctive type; they are a concern of this study only where they have been employed within public-facing genres, to explain or to extend the visual palette of filmmakers.

The book's original subtitle was 'Paul Rotha and the Definition of Scientific Documentary'. The final book honours its intention both by reflecting on the different definitions used by the historical actors and by constructing its narrative around the career of the documentarist Paul Rotha, who acts as Virgil to my Dante, leading me through the unknown territory of twentieth-century science films. This device resonates with the volume's use of participants' categories. Rotha was the most prolific of British documentarist writers and a considerable filmmaker too (see chapter two). This account travels with him as he learns techniques of montage and uses them to represent the modernists' world of applied science and technology. We note his extended networks of scientists, planners and filmmakers who collaborated to produce new representations of this world. Rotha's developing beliefs about the proper representation of the scientific and technological world, of a piece with the application of rationality and planning more generally, provide the viewpoint from which other scientific films are seen. Rotha's biographical trajectory takes us into the heady days of television's determining decade, the 1950s. In brief, the story of Rotha's engagement with the representation of scientific technique dominates the core of this book. But such an account risks being too particular; accordingly this study looks at the scientific films for laypeople made before his period of activity. It also considers how the world of scientific filmmaking changed after his period of influence.

Rotha is not generally considered a scientific filmmaker. In the chapters that follow, I select and stress Rotha's concerns with science. If this book were Rotha's biography, this would be a distortion, but it is an account of scientific films and Rotha was, on the evidence amassed here, indisputably a significant maker of films that represented science, technology and medicine. It is also clear that Rotha's filmmaking practice and social beliefs were a developing, sometimes contradictory, whole. But, just as there are many potential histories of scientific films, so there are several potential histories of Paul Rotha.

A focus on a single filmmaker's work should not be taken to signal an auteurist interpretation (see Cook 1985: 114–206). Films could not be made without the financial, practical and intellectual contributions of people other than film directors, and such people often had a major influence on how subjects were represented. Powerful institutions speaking for the interests of science, such as the Royal Society or the British Association for the Advancement of Science, have often had an influence on the representation of their subject, as have science- and technology-based industries with public relations budgets. But that is not to say that filmmakers have been unwilling participants in the representation of science; for a variety of reasons that the individual chapters discuss, those that represent science and technology have usually had commitments to the significance or value of science. In the concrete world where films cost money, they have generally only been made when it has served the interests of powerful groups within society able to fund such activities. There is always an economic base and that base has always been tied to interests.

Authorship of films can be seen to extend beyond the director towards the whole range of people involved in each film. Rotha (who saw editing as intrinsic to directing) expressed something similar in 1933: 'only one man can write a script of a film and that is the director and the cutter ... But that does not include content. The subject, theme and even narrative situations can be suggested (not written) by outside sources and adapted as thought fit.'[2] Most often, these associations between filmmakers and others have involved shared ideologies of one kind or another. They were sometimes explicitly political; many of those who worked in documentary were on the left, without necessarily being members of any particular group or party. Rotha was typical of this, positively refusing to be categorised any more specifically than as a socialist, yet happily reading Raymond Postgate's *Karl Marx* (1933) and discussing the Marxist interpretation of history. At key points in the chronicle of the visual representation of science, scientific filmmaking has thrived because networks of scientists and filmmakers have agreed to the importance of science for extra-scientific and extra-cinematic reasons. This book is written on the assumption that films were the products of such extended social networks, but I have not laboured the point here.[3] My approach, stressing relationships between people, is closer to a prosopographical analysis (though with a stronger emphasis on short- and long-term associations between individuals than an exhaustive analysis of shared social origins) than to Latourian actor-network theory (see Stone 1971; Latour 1987).

Films of Fact is a history of production; it is concerned with individuals and institutions from, on one side, science and, on another, filmmaking and broadcasting. The public complete the triangle. But, despite the importance of audiences, this book does not extend far into questions of reception. Steve Shapin argues that science has been kept separate from the public only as a result of 'the enormous labour expended by individuals in the past in constructing the very categories of "science" and "the public" and in stipulating the proper nature of transactions between them'. Furthermore, these categories have been contested: 'different groups of interested persons upheld divergent views of what science was or ought to be and how the boundaries between it, other forms of knowledge and public concerns ought to be drawn' (1990: 991, 992). As historians have noted, 'even when it is clear that popularisers of science sought to inculcate middle-class values and scientific optimism, it does not necessarily follow that science in popular culture functioned to that end' (Cooter & Pumfrey 1994: 247). Shapin therefore recommends study of the 'vehicles used to communicate between science and the public … because the conventions and distribution of these vehicles may have had an important bearing upon public perceptions of scientific claims' (1990: 1000–1). Following this argument, and accepting the importance of audiences, I have opted to study only the vehicles of scientific communication, and not to pursue the serious questions of audiences merely at the periphery of the book. Future works will fully integrate discussion of audiences with the production of films, but those works, radically different in kind, and demanding significantly different research techniques from those in the current study, await further attention.

1 SCIENCE, NATURE AND FILMMAKING

> It will be through the perseverance of the scientist who has the happy faculty of amusing as well as entertaining a general audience that this class of film ever will have a vogue. The tendency towards this state of affairs, happily, is increasing every day, for the cinematograph is appealing more and more to the cultured classes, who, after all, constitute its most substantial support. (Talbot 1912: 196)

The one-minute film *Cheese Mites* was the sensation of the first public programme of scientific films shown at the Alhambra Music Hall in Leicester Square, London, in August 1903.[1] The film's eponymous subjects were revealed to the astonished spectators by means of the show's novelty – a motion picture film shot through a microscope – as 'crawling and creeping about in all directions, looking like great uncanny crabs, bristling with long spiny hairs and legs' (Charles Urban quoted in Gaycken 2005: 60). To amplify the effect, Urban, the businessman and showman behind the event, soon added an introductory shot showing a man eating his lunch and detecting that there is something the matter with his Stilton cheese. 'He examines it with his magnifying glass, starts up and flings the cheese away' (McKernan 2003: 73).

This first Alhambra programme also included the microcinematographic studies *The Frog, His Webbed Foot, And the Circulation of his Blood; The Fresh Water Hydra*; and *The Circulation of the Protoplasm of the Canadian Waterweed.* These were alternated with the observational and moralising animal behaviour films *The Greedy American Toad; The Pugilistic Toads and the Tortoise Referee; Chameleons Climbing and Feeding*; and *The Boa Constrictor (six views)*. The second part of the programme was made up of 15 short films on bee culture 'by courtesy of Mr C. T. Overton, Bee expert, Crawley' (Anon 1903c; see also Gaycken 2005: 47, 70).

This was almost certainly the earliest occasion on which a British lay audience was shown scientific films, but these were far from being the first scientific films to be made. Indeed, most histories now ascribe the origins of cinematography to several different scientific and technological sources, notably to the entrepreneur-inventor tradition of Thomas Edison, the scientific activity of Étienne Jules Marey's chronophotographic studies, and innovation in photographic technique exemplified by Eadweard Muybridge, who worked in collaboration with a railway engineer (see Christie 1994; McKernan 2004b; Tosi 2005).

After the Lumière Brothers demonstrated the technology in 1895, doctors and scientists very soon began to experiment with it, applying it as an instrument for scientific research. L. Braun filmed the mammalian heart, Paul Schuster produced studies of patients with conditions including Parkinson's disease, Robert Watkins made microcinematographic renderings of blood corpuscles, John Macintyre – a consultant Ear, Nose and Throat surgeon – created a stop-motion X-ray film of a frog's legs flexing, and the surgeon Dr Eugène-Louis Doyen started making surgical films (see Essex-Lopresti 1997: 819–20). Most of the activity was in the medical sciences, but there was also a small amount of production in anthropology and astronomy too. For doctors and scientists, to employ the cinematograph was to continue within the culture of visualisation that had contributed to its invention in the first place. Hannah Landecker has suggested that microcinematography had a revolutionary potential in the sciences, especially as it made visible phenomena that were too slow or too fast to be perceptible (2006: 123). This potential was constantly asserted across the twentieth century by the advocates of scientific experimentation using the cine-camera.

None of these films, made by doctors for the medical lecture theatre or the scientific meeting, was intended to be seen by the general public.[2] They were certainly part of the new visualising armoury of the sciences, but they were not part of its public culture. Yet the show that the audience at the Alhambra witnessed was not of experimental films made by scientists in their research; the programme that featured *Cheese Mites* was of scientific films specifically intended for the general public. The programme, billed in full as *The Unseen World: Revealing Nature's Closest Secrets by Means of the Urban-Duncan Micro-Bioscope*, was comprised of a sequence of short films running for about twenty minutes in total. Charles Urban here presented the work of his new employee, Francis Martin Duncan, whom he had engaged to provide microcinematographic sequences as parts of the programmes of topographical and scientific 'actualities', in which his new company was established to specialise. Duncan, like popular magic lantern operators before him, gave a lecturer's accompaniment, in a style described as 'concise and cheery' (see Gaycken 2005: 40, 47, 55).

The showing of *Cheese Mites* marks our first cinematic encounter in the highly variegated and ambiguous territory of science and technology in public culture. This region is bounded on one side by certain varieties of élite science and on the other by the shading-off – by infinite gradations – into the technical knowledges of lay people. This ground is contested and some areas within it, the Alsace-Lorraines of science, have changed hands many times, sometimes considered 'scientific', sometimes 'lay' or 'amateur'. Since the 1970s, writers in several disciplines, including practitioners of the sociology and social history of science, and advocates of the public understanding of science, have mapped this territory in different ways. It has tended to be seen not as a whole, but as a series of different issues, and often of agonistic relationships: between knowledge and ignorance, élites and masses, professionals and amateurs. In that sense, the realm of science is like a scale map of the wider territory of general social history. And so it is little surprise that, as far as many of these writers are concerned, the development of science has exemplified the exercise of social power in modern societies. One thing is clear from all this activity: that a binary separation of 'science' on the one hand and 'the public' on the other cannot do justice to the variety of activities and beliefs of the inhabitants of this zone. Between the swampy margins of this territory, every space is occupied by some type of activity that is scientific or technological to some degree. Although historians have studied this terrain, it is not such well-trodden ground that there is a single obvious path to follow in the analysis of early scientific films.

In *The Unseen World* and its ilk several different kinds of scientific and technological activity coincided: the science shown (it is significant that cheese mites are a living natural phenomenon), the 'scientific' professional identity of filmmakers, and the apparatus they used. This represents a complex of issues that can only be understood in the context of how the sciences and technology had developed in the nineteenth century. It is necessary to understand what kind of activity the invention and subsequent development of the cinematograph represented, not least because businessmen and filmmakers, including Charles Urban, played on the scientific status of cinema in their appeal to the public. It will be seen that an integral part of our discussion must be the relationship between élite and popular science, and the nature of popularisation and public culture.

Changes in the sciences

The nineteenth century witnessed a revolution in the sciences in which the relatively small-scale eighteenth-century disciplines of natural history (which recorded and classified the world) and natural philosophy (which explained and investigated natural phenomena) gave way to large, influential and powerful sciences. The older disciplines, over the course of the century and at different rates, were succeeded by large-scale specialised, professionalised and distinct disciplines such as physics, chemistry, mathematics, biology, and earth and social sciences which, by 1900, had acquired the more or less stable forms that they have retained (see Cahan 2003a).[3] A similar pattern can be seen in technology, as the age in which a single engineer might assay the design of ships, engines, bridges and other structures, was superseded by an era when specialist varieties of technological expertise and engineering were developed – each with its own codes of practice, bodies of technical knowledge and standards of precision (see Wengenroth 2003: 243–6).

To elucidate the scientific world into which cinema emerged and developed, we may allude to John Pickstone's 'ways of knowing' model of the history of science. He represents the field of science as a layering of three 'ideal types' of activity, each of which has had a period of greatest currency in the past, but which may be found with varying degrees of concentration throughout the last three hundred years. The first of these is an observational and descriptive natural-historical approach to 'nature', the world that the sciences represent (this is not necessarily equivalent to 'natural history' in the now more common sense, denoting the study of living things). This was the dominant mode of the eighteenth century, but continues as a foundation of many sciences. The Urban-Duncan films are closest to this mode. The second type is an analytical mode that quantifies and breaks down its objects of study into their component parts, for example the sodium and chloride of salt, or the bodily tissues distinguished by histologists. This mode had its first heyday in the sciences in the period following the French Revolution. The third type is an experimental approach in which procedures are developed that – complementarily to analysis, which takes apart – puts together elements to create new phenomena. Here he gives the example of the 'control experiment', defined in the work of the physiologist Claude Bernard, in which all elements except the one under investigation are retained constant. This mode flourished particularly in the last third of the nineteenth century. Many teaching films reproduce both the analytical and the experimental mode. To these three ideal types, Pickstone adds a mode in combination, which he names 'technoscience', the production of scientific commodities in 'academic-industrial-governmental complexes', of which the post-World War Two industrial modes of research are the most obvious (see Pickstone 2000: 8–15). This complex mode provided the context for most of the films discussed in later chapters.

The popularisation of science

As the new sciences grew in size and reach, their links with other aspects of culture also multiplied (see Cahan 2003a: 3–15). At the same time as the scientific professions became more specialised, some also sought to forge new relationships with the public. This was in the activity called popularisation. Roger Cooter and Stephen Pumfrey commented in their analysis of science in popular culture that

> there are obvious differences between the study of 'popular' science ... and the study of the myriad processes of 'popularisation'; just as there are significant differences between these enterprises and the history of 'populist science', 'proletarian science' and what has been styled 'pop science'. (1994: 239)

They reach the conclusion that, if we choose to use the terminology of popularisation, 'the very language we use ... belongs to a discourse of analysis that is ideologically and culturally loaded: as it assumes differences or boundaries between cultures and classes, so it assumes that the only object of analysis is the transfer of cultural items across such boundaries' (1994: 248). We might add that it also obscures the richness of the middle ground. But they bring out the ideological dimension: where scientists have emphasised the esoteric nature of scientific knowledge, this has served to maintain the distance between them and the public (1994: 240).

The action of the verb 'to popularise' – used here in the sense of seeking to render élite science popular – is traditionally concerned with passage from high to low, from complex to simple, from knowledge makers to knowledge consumers, from active to passive. The main medium from the eighteenth century onwards for the popularisation of science in this sense was publication. After the introduction of rotary printing presses from the 1860s made it possible to publish cheap editions, there was a significant vogue for popular science books (see Bowler 2006). In the twentieth century, some authors, such as J. B. S. Haldane, were practising scientists, drawing salaries from universities or similar organisations. Others, such as 'Professor' A. M. Low, author of *Thanks to Inventors* (1954), were members of the scientific community only tangentially, and their income depended on their popularising activities. Many science writers were generalist journalists (see Cantor *et al.* 2004) but, from the 1920s, some became science specialists, notably Gerald Crowther and Peter Ritchie Calder, whom we shall encounter in later chapters. And, in those styles of popular science conducted by non-professionals, some people made the transition from 'popular scientist' to published author, although their productions were often intended first for their own communities. One example is the artisan botanist L. H. Grindon, who published his *Manchester Walks and Wild Flowers* in 1859 (see Secord 1994: 272).

But the intended audiences for these books and articles may often have possessed more *scientific* knowledge than their authors conceded. And many in the 'lay' audience also possessed considerable *technical* knowledge, often derived from their experience of trades. Furthermore, any professional scientists seeking control over what was represented as science in public culture had to compete in a highly variegated marketplace. It is clear that, between the world of the élite scientists and the lay consumer or practitioner of science, there has existed a diverse subculture of middlemen and women with expertise claims that have varied over time.

The visual and participatory culture of science

Representing or embodying science, the Urban-Duncan films were part of the visual and participatory cultures of science, technology and medicine that was richer, more ambiguous and less intrinsically hierarchical than popularisation. This field has been sufficiently explored by scholars that one of its practitioners is able to allude to 'a new historiography of visual culture' (Morus 2006: 107).[4] Iwan Morus comments that 'there was a proliferation of [natural] philosophical technologies of display during the latter half of the eighteenth and the first half of the nineteenth century. Audiences flocked to see electrical exhibitions, magic lantern shows, phantasmagorias, and similar wonders.' These phenomena, as he states, could be experienced in cities across Britain, Europe and North America (2006: 102, 105). From the mid-nineteenth century, world exhibitions – notably the Great Exhibition of 1851 and that in Paris in 1900 – with their halls of machines and manufactured goods, set precedents for public experience of the products and wonders of the industrialising nations. Specialist exhibitions mixed displays in the mode of the 'natural history' way of knowing with others which stressed the spectacular. The International Health Exhibition, held in 1884 on the South Kensington site previously occupied by the Great Exhibition, was one example (see Adams 1996: 9–35). The second half of the nineteenth century was also a boom time for public museums, including many devoted to the sciences and technology, especially the South Kensington Museum, founded in 1857, which contained exhibits of technological history, decorative arts and scientific apparatus.

But much of the visual and participatory culture of science was to be found on a smaller scale, in visual experiences and performances of many kinds. The magic lantern played a significant role in the public culture of science. One of its showmen, Henry Pepper, promoted its wonders by combining its instructional capacities with its power to amaze:

> Its educational importance is now being thoroughly appreciated, not only on account of the size of the diagrams that may be represented on the disc,

but also from the fact that the attention of an audience is better secured in a room where the only object available is the diagram under explanation. (Chanan 1996: 14)

The lanternists lecturing about science not only had a mediating role between 'the general audience and the nineteenth century gentleman amateur' (ibid.), they also acted as stimulants for inventors in the sense that they took up improvements to the lantern as a means to enhance the attractiveness of their performances. Amongst such innovations were those described by Charles Musser, who gives the example of a magic lantern in 1866 adapted to project through a miniature aquarium containing fish, and also of hollow slides containing live insects whose movement could thereby be projected onto the screen (1990: 32; see also Gaycken 2005: 17). Spectacular shows of living things were evidently a significant component of the entertainment culture of the period.

Iwan Morus cautions against the teleology of presenting the rich visual culture of popular science as merely an antecedent to cinema. We need to appreciate, he argues, 'the ways that these optical spectacles were experienced by their audiences, who presumably regarded them as fully fleshed-out experiences in their own right rather than as milestones on the way to twentieth-century cinema' (2006: 104). Any full account of cinema's origins would need to consider not only the technical and economic aspects of the invention, but also to appreciate the ways in which nineteenth-century cultural genres and popular entertainment played a role, both in constituting the forms taken by the first films and also in establishing the attitudes and expectations of audiences encountering them (Chanan 1996: 7). In the spirit of Morus's caveat it is wise also to avoid whiggishly viewing early cinema as being 'on the way to' the cinema of many kinds that subsequently developed. To gain a historical understanding of the audience experience of a new form of entertainment it is necessary to focus on the specific context and to understand that spectators had to learn how to see novel phenomena (2006: 107).

Part of the attraction to audiences was what Tom Gunning has called their 'aesthetic of astonishment'. It is important, as he has shown, not to give credence to the foundation myth of cinema that spectators at the first showing of the Lumières' *Arrival of a Train at the Station* (*L'Arrivée d'un train à La Ciotat*, 1896) were experiencing something entirely novel or that they were terrified by the realism of the experience. Instead of assuming the audience to be naïve and credulous, we should understand that 'the projection of the first moving images stands at the climax of a period of intense development in visual entertainments, a tradition in which realism was valued for its uncanny effects' (2004: 864). Rather than mistaking the image for reality, he suggests that the spectator, in 'a vacillation between belief and incredulity', enjoyed the new quality of the illusion that the cinematograph supplied (2004: 867). He explains how the showman projectionist in the early Lumière exhibitions would start the films on a still frame then begin to crank the projector, bringing the still picture 'to life'. The providers and consumers of a wide range of nineteenth-century visual entertainments shared in this 'aesthetic of astonishment' (2004: 866–7).

The cinema emerged into an existing and vibrant entertainment culture in which scientific novelties were an established genre. The *Morning Post* suggested of the Urban shows at the Alhambra that 'as a music hall turn, last night's production was an unqualified success', pointing to the context in which early films were often shown. The fact that this 'turn' was later promoted to top of the bill shows that the films were shown in the context of the usual miscellaneous entertainments of the Music Hall. A listing for the Alhambra in late August makes the point:

> New Grand Dramatic Ballet/CARMEN/The Legendary Ballet THE DEVIL's FORGE/Le Roy, Talma and Bosco (The World's Monarchs of Magic), in their Wonderful Conjuring and Illusions. The Unseen World on the Bioscope, Exhibited by the Charles Urban Trading Company. Grand Varieties. (Anon. 1903e: 18)

The first British Lumière programme in 1896 had similarly shared a bill with, amongst others, 'Belloni and the Bicycling Cockatoo', whereas in 1910 a sequence of Kinemacolor films was the culmination of a programme at the Palace Theatre that also included the acts 'Albert Whelan, The Australian Entertainer' and 'Vickey Delmar, Novelty Dancer' (see Chanan 1996: 127, 26).

Natural History subjects

The fact that the predominant subjects of the early scientific public cinema were natural historical is no coincidence. The history of these sciences, with their existing pattern of widespread and active popular participation, gave them a place in the general culture that made them likely subjects for the first scientific films. Nineteenth-century promoters of lay involvement in natural history had promoted a 'bucolic' view of the subject as pleasant and useful, both by choice of subjects – including floral display and animal behaviour – and by portraying the naturalist as an adventurer. Visual media, including illustrations and latterly films, were 'crucial to this strategy' (Drouin & Bensaude-Vincent 1996: 410). Natural history was encouraged variously as fostering a Romantic notion of aesthetics, a stimulant to the growth of reason, a succour to religion and a source of aesthetic pleasure. Amateurs might enjoy natural history as natural theology, to wonder at the beneficence of the Creator, or later at the diversity of the world that natural selection had wrought.

There had been deliberate efforts from the time of Georges Cuvier (1796–1832) onwards to separate the natural historical interests of the general public from those of professional practitioners for the ostensible reason of the technical difficulty of nomenclature and taxonomy. But, on the ground of nineteenth-century natural history, it is difficult to establish a clear distinction between professional and lay natural historical activity or to clearly divide 'active' participants from a passive 'general public'. Active natural historians included wealthy gentlemen specialists, university and museum professionals, and amateurs with significant collections, all of whom were not only consuming but doing natural history (ibid.).[5]

Natural historical knowledge cultures also played a role in the self-definition of artisans; these non-élite groups 'were actively involved in making their own intellectual world' (Cooter & Pumfrey 1994: 242). Anne Secord argues of nineteenth-century artisan botany that 'the artisan mentality' depended both on independence, which included a concern with social status, and on the possession of skill. The pursuit of botany operated within an idea of skill as a type of property; skill in collecting, identifying and classifying plants was of a kind with their skills as craftsmen, and served to reinforce their class identity, not least when merchant capitalism served to deskill groups such as handloom weavers (see Secord 1994: 292).

The literature on the social location of science tends to treat the classes separately and with different terminologies, so that similar activities of collecting, identifying and classifying, when chosen by artisans are called 'popular science', and when enjoyed by the middle classes or aristocrats are called 'amateur science'. For the individual, the difference in the meaning of the activity may well have had a great deal to do with their particular class-consciousness in their particular historical moment. But processes of class definition also worked by deliberate exclusion:

> The contest over science in the early nineteenth century had been one about who could participate and on what terms. The *result* of this contest was ... the redefinition of popular science. Fearful of the ability of the working class to appropriate knowledge for its own ends, the middle class increasingly rendered working-class scientific activity politically neutral through control over printed texts ... producing accounts of the lives of autodidacts to put forward moral lessons, and by giving natural history a central role in rational recreation. (Secord 1994: 299)

Secord comments that 'increasingly ... the term "popular science" was used by the dominant culture to signify bodies of literature and scientific activity that had little or no interaction with élite science' (1994: 297). But it would be a highly reductive and ideological reading to claim that the middle classes invented the new forms of science with the *sole* aim of advancing their class interests. And, in fact, the processes by which artisan participants were excluded from professionalising science also marginalised 'clergymen, women and "non professionals" in general, thereby producing the category of the "amateur" whose work was rarely of use to the professional scientist' (ibid.). The significance of this category is emphasised by the 'semantic switch' in the term 'amateur' in the last decades of the century: 'the old positive meaning of "connoisseur" has gradually been overthrown by the pejorative sense of "dilettante" emphasising a lack of seriousness and reliability' (Drouin & Bensaude-Vincent 1996: 417–18).

But, as in the wider picture, by the time of the Alhambra shows a revolution was well under way in the biological sciences. From the 1870s, the laboratory 'way of knowing' was being elevated over the older natural historical approach that was dominated by the gathering, identification and classification of specimens (see Nyhart 1996: 439–42). The process of 'defining the laboratory and the experimental station as the sites of legitimation of botany and zoology' (Secord 1994: 297) marginalised lay involvement. That forced lay practitioners, of whatever class, into the role of amateurs or at best collectors on behalf of salaried scholars, who kept to themselves the expert practices of analysis and classification. But, 'even when natural history was being denigrated by exponents of analysis or experimentation, it remained a major element of *popular* science, an important means by which professional biologists could extend their work and influence, and a key to public understandings of "nature" – as indeed it remains' (Pickstone 2000: 76). This was the twilight world in which Duncan and his natural history filmmaking successors operated: excluded from élite biological science and finding common cause with the artisan classes, which had often been the intended audiences for improving entertainments.

The style of science in *The Unseen World*

An important part of Charles Urban's claim to be producing scientific films rested on the qualifications of his staff; his manifesto, *The Cinematograph in Science, Education and Matters of State*, described the films as the 'results of the labours of trained and qualified scientific experts' (1907: 10). This referred both to the 'scientific' activity of visualising microscopic and other life forms, and to the technological means to achieve this. From what can

"THE UNSEEN WORLD"

(Copyright Title.)

A Series of Microscopic Studies, Photographed by means of

The Urban-Duncan Micro-Bioscope

The "UNSEEN WORLD" Series of Films are made to fit all Standard American Gauge Projecting Machines.

The magnification of these Subjects as viewed from a Screen, with picture 20 by 25 feet in size, is 2,200,000 to 28,000,000 times, according to the extent of magnification on the Film which varies from 25 to 850 diameters.

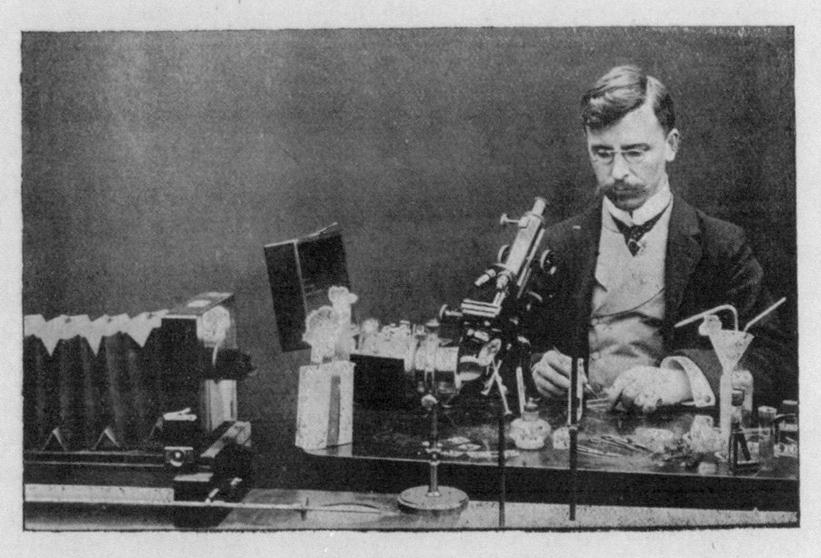

F. MARTIN-DUNCAN, F.R.H.S.

Fig 1.1 Francis Martin Duncan in Charles Urban's Catalogue (1905)

be learned of his staff in this period, Francis Martin Duncan, who shared the honour of the naming of the apparatus – the 'Urban-Duncan Micro-Bioscope' – was the individual to whom the label would best apply. Urban's language in the phrase 'trained and qualified scientific experts' nods to the development of professionalised science. But in what ways might contemporaries have considered Duncan a scientist? Much depends, of course, on who might have been making the judgement.

Jack Morrell writes of professionalisation – a significant element of the nineteenth-century revolution in the sciences and technology – as a form of occupational development and strategy of scientists. In this, 'the desire for higher status, autonomous control of conditions of work, and control of the market in the interests of higher rewards (financial and honorary), were all prominent' (1990: 981). In his account, most professionalised sciences combined several factors: an increase of paid positions and specialist qualifications linked to the possession of scientific knowledge; the development of laboratories into spaces for the acquisition of experimental skills; specialisation and subdivision within the sciences (with a proliferation of technical languages); growing group consciousness and solidarity; and reward systems including the honorific fellowship of scientific societies, such as the Royal Society (see Morrell 1990: 981–4). In 1900, whatever the dynamics of the preceding decades, to be a scientist was generally associated with salaried work, most often in universities (see Cahan 2003b: 301). Also, for most non-professionals practising the sciences, there now existed related professions against which they might compare themselves – or be (invidiously) compared.

Duncan was not a professional experimental biologist. Rather, he was a practitioner at the intersection of amateur natural history and photography, as is evident from his handful of publications up to the time he left Urban's employment in 1908. According to the *Amateur Photographer* in 1903, although 'still a young man … his experience extends over fifteen years … and he has acquired a knowledge of the subject [photomicrography] that few men possess' (Anon. 1903a: 176). During the 1890s, he had explored moving pictures by mounting still photographs of micro-organisms in a zoetrope (see McKernan 2003: 69–70). The year before taking up employment with Urban, he had published a handbook on photomicrography in a series for amateur photographers (Duncan 1902). His first published nature photographs had appeared in the six-volume *Cassell's Natural History*, edited by his father, the amateur naturalist Peter Martin Duncan.[6] In 1900 Francis published *Wonders of the Sea*, a series of children's natural history books. We find him attending the Camera Club and publishing short technical articles in the *Amateur Photographer* on such subjects as 'worms, and how to photograph them' (1903). This example gives a clear impression of the style of his practice. His lengthy description of how to build a camera stand and a tank for photographing marine worms reveals the tinkering amateur technologist: 'all the pieces of glass must be thoroughly cleaned … then warmed and fixed together with the best marine glue'. Similarly his is an enthusiast's engagement with photographic technology; he expatiates on preferred camera format (quarter plate) and lens ('my own particular "fancy" is Messrs. Watson's Holostigmat Convertible'). Although Duncan knew his natural history well enough to cite Darwin, he used the same lay enthusiast's tone for the creatures themselves: '[worms] are shunned and avoided, and generally voted to be "nasty slimy things". And yet, when you get to know them, and understand their ways, the worms, like so many humble and despised creatures, are not only interesting and entertaining, but are beautiful withal' (1903: 48). The piece's conclusion, which provides advice on collecting, is full of the amateur naturalist's aesthetic delight: 'The play of iridescent colours on the hairs which cover the back and sides of the sea mouse ["that beautiful and most unwormlike worm"] is most exquisite, all the hues of the rainbow play along them with every change of light' (1903: 49).

Professionalisation differentiated professional from amateur, a distinction that was often socially neutral in that the amateur scientist may often have been of the same or of higher social status than the professional. Francis Martin Duncan's father, for example, earned his living as a physician whilst he became an eminent amateur geologist with many scientific honours (see Bonney & Foote 2004). But professionalisation also differentiated profession from trade, and this was effectively code for social class. The Victorian middle classes ascribed value to belonging to a profession, which played to contemporary concerns with status and respectability: 'professional people dealt with people *qua* people, while tradesmen and artisans provided things for their physical wants'. Proto-professionals in the sciences sought the status of the long-established professions of divinity, law and physic (Morrell 1990: 980). Duncan, middle class by virtue of being the son of a physician, in his microcinematography of natural historical specimens was operating as a gentleman amateur. Equally, natural historical pursuits could also play a significant role in the self-definition of the landowning classes (see Carroll 2004).

Historians have as yet uncovered no reason for Duncan's departure from Charles Urban's company in 1908. But we may speculate that employment at the Zoological Society offered a professional scientific association and a class status not available in the liminal commercial world of the Music Hall non-fiction specialist. Although his original post is unknown, he was the Zoo's Librarian between 1919 and 1939 (see Fish 1976: 249).

Duncan's successor as Urban's 'scientific cinematographer' was a clerk from the Board of Education, Frank Percy Smith. He too was an amateur naturalist with considerable inventive skills in the amateur tradition. Amongst Smith's more than fifty films for Urban were *The Balancing Bluebottle* (1908), which showed a fly juggling a wine cork much larger than itself, several

Fig 1.2 Frank Percy Smith at his 'studio' at Southgate, London

microcinematographic studies and the time-lapse colour film, *The Birth of a Flower* (1910) (see McKernan 2004a).

We may better locate the kind of science the Duncan and Smith films represented by paying some attention to the style of their 'scientific' activities. Both men were, for longer or shorter periods, members of the Quekett

Microscopical Club (QMC), which met in London. The term 'club' tells us a great deal about the organisation. At its foundation in 1865 it was established with the aim of 'affording to microscopists ... opportunities for meeting and exchanging ideas without that diffidence and constraint which an amateur naturally feels when discussing scientific subjects in the presence of professional men'. This comment referred to meetings at the older Royal Microscopical Society. The stress of the QMC was decidedly on amateur activity; in fact the letter in the magazine *Science Gossip* that launched the club stated that it would be run 'on the plan of the Society of Amateur Botanists'. This is 'amateur' in the older sense of 'connoisseur', with a nod to the cultures of popular botanical science explored by Anne Secord. The Club's prospectus conveys its context within scientific popular culture: 'The increased study of natural history in late years has created a large class of observers, who, although with limited leisure for such pursuits, possess notwithstanding, earnestness of purpose and ability to render good service in the cause of microscopical investigation' (Jones 1949: 1). Its business was conducted via the panoply of amateur club activities: monthly meetings, 'gossip nights', excursions, the founding of a library and slide collection. That ethos continued in the vein of its founders at least until the admission of women in 1918 (see Spitta 1923; Barron 1965). The membership included a cadre of individuals with a miscellany of amateur interests and pursuits, for example Edward Milles Nelson, president in the 1890s, who liked to describe himself as an 'amateur dabbler', expert in photomicrography and 'life-long contributor to the *English Mechanic* on microscopy, astronomy, mathematics, etc.' (Barron 1965: 66).

Frank Percy Smith's involvement in the Club was significant, as he was a member from 1899 until his death in 1945, and the editor of the *Journal of the Quekett Microscopical Club* from 1904 for five years. Between 1904 and 1909 he contributed seven articles on different species of spiders, two classifying specimens of spiders and one on 'mounting spider dissections as microscopical objects' (listed in Newstead 1994: 179). His paper on 'The British Spiders of the Genus *Lycosa*' is typical both of Smith and of the amateur scientific discourse of the Club. He aimed, so as to ease the facility with which a beginner may identify a particular spider, to provide 'a concise comparison of the various species with the descriptions limited to characters which are not common to the whole genus' (Smith 1907: 10). He helpfully remarked that 'the drawings, except where otherwise stated, are from spiders in my own collection' (1907: 13). The bulk of the article was a minute description of the 18 native British species. The discourse was descriptive and taxonomic, aimed at other amateurs who might wish to identify spiders. The expertise of the amateur collector in finding specimens is conveyed in a vignette of 'initiating a friend into the subtleties of spider-collecting … during an autumn ramble' (Smith 1908: 312).

Duncan seems to have been a member of the Club for a much shorter period, and over a decade after his employment by Urban, but the discourse of his contributions was similar to Smith's. He gave talks on 'studies in marine zoology' in 1919, 'some aspects of the Crustacea' in 1922, and on 'insects as disease transmitting agents' in 1918. His largest substantial paper was a technical account of 'some methods of preparing marine specimens' (Duncan 1921). The parts of nature may have been different and the prose more mellifluous ('In few branches of Natural Science has the microscope been of greater service than in the study of minute plant and vegetable life that swarm in the sea' (1921: 215), but the enterprise denoted in these studies was the same as Smith's; essentially that of the experienced amateur naturalist. He wrote, for example, that 'a morning's ramble along the seashore after a heavy gale … will often supply sufficient material to keep the most ardent microscopist busy for many a long winter evening' (1921: 215). He too wanted to stimulate beginners' interest by conveying techniques he had developed over 25 years. He recommended a half-gallon pickle jar to keep the catch. Not only is Duncan's account similar to Smith's, but it was broadly typical of the contents of the *Journal of the Quekett Microscopical Club*, with its microscopically investigated description of genera and explanations of technique.

Technology and science: invention

Duncan and Smith fused amateur expertise in natural history with specialised photographic expertise allied to a 'tinkering' capacity to make new apparatus that was also typical of the nineteenth century. Non-professionalised styles of innovation had created the cinematograph and, later, the modifications that permitted the creation of the sensational *The Unseen World* films. This is shown in Urban's discussion of the technical difficulty of filming bacteria. The problem is set out in seemingly scientific language, referring to 'the close affinity of [the] refractive index of [living bacteria] to the media in which they are cultivated' (1907: 41). But the solution was pure tinkering: 'after months of experiment with various optical formulae, our scientific staff succeeded in finding a combination of lenses which permit accurate examination of living, unstained bacteria' (ibid.).

Bernard Carlson states that 'creative technologists', throughout the commercial and industrial revolutions, had 'styled themselves inventors' and that they 'frequently began as craftsmen, and their ability to envision and fashion new machines was grounded in their profound understanding of craft' (1997: 205). The style of innovation was that of the tinkering artisan fashioning new machines by increment and combination of existing mechanical principles and devices (see 1997: 204–8). Invention in this mode was not a matter of 'a sudden flash of inspiration from which a new device emerges "ready made". Largely it is a matter of minute and painstaking modification of existing technology' (Mackenzie & Wajcman 1999: 9–10). Following Harold Perkin's discussion of expertise and Anne Secord's analysis of artisanal knowledge culture, we may see this skilled manual expertise as a type of human capital that the inventor sold to their employer (see Perkin 1989: 378; see also chapter two). This was the style of innovation within which the pioneers of cinematography worked.

The more general transition from this 'tinkering' mode of invention first became clear in the context of the big technological businesses of nineteenth- and twentieth-century America. First, seeing the competitive advantage of innovation, they employed inventors on the old model. But later, industrial concerns competing to produce similar goods began to squeeze out the self-styled artisanal inventor in favour of centralised research facilities and university-educated scientists. 'Managers shifted from inventors to scientists in part because the image and rhetoric of science appealed to managers intent on protecting their organisations by minimising risk. Inventors fell by the wayside because they came to be perceived by managers as an unpredictable source of innovation' (Carlson 1997: 223).[7] This was the beginning of the reign of 'applied science' in distinction from artisanal invention.

It is important to unfetter our conception of the cinematograph that emerged into public view in 1895 from the assumption that it was either the product of a 'flash of inspiration' or yet merely the *end* of a process of invention. Its ambiguity is borne out by the fact that, like the X-ray set, its contemporary visualising invention, it did not immediately resolve into a single perceived category or dominant pattern of use (see Harding & Popple 1996: 18). Was it fated merely to be 'a superior magic lantern combining motion with pictorial effect' as a writer in *Chambers Encyclopaedia* in 1900 described it? (Whitby 1996: 21). These were questions of definition. What did it mean to make or show a film? What should it be? What could it be? Was it an entertainment or a scientific tool? Did it belong principally in the laboratory, the fairground or elsewhere? Urban's success with the films from *The Unseen World* was significant in establishing the market conditions for a later range of science programming well beyond the range of the micro-bioscope.

Urban's definitions

Charles Urban's business relied on creating new markets and he was as intent on creating 'serious' use in education, science and business as he was on entertainment. And yet the evidence is that when Urban sought to recommend films for more sober employment, he used as examples the same films and types of film as were seen at the Alhambra shows and their successors. Urban's most extended claims to be making and exhibiting scientific films appear in his 1907 manifesto, *The Cinematograph in Science, Education and Matters of State*, which argued for serious use. With the showman's disregard for inconvenient distinctions, he argued that the Alhambra shows had a 'purpose in attracting and compelling the attention of scientists and experts' (1907: 7), when there is little evidence that these were significant audiences. But Urban asserted that the success of the enterprise had resulted in 'prepared educational and scientific series of subjects' (ibid.).

> The time has now arrived when the equipment of every hospital, scientific laboratory, technical institute, college and private and public school is as incomplete without its moving picture apparatus as it would be without its clinical instruments, test tubes, lathes, globes, or maps. (1907: 9)

Whatever the argument for the use of films in the classroom – and these continued for at least thirty years (see Low 1979a: 13–17) – Urban's claim for the potential of the cinematograph as an instrument of scientific and medical research is striking. But, if this was a bid to make research films for scientists, then there is no sign that he was successful. The Commission on Educational and Cultural Films reported in 1932 that 'not much has yet been done in Great Britain' in the use of the cinema for scientific investigation (Anon. 1932: 116). And, if Urban was voicing a will to popularise élite science, then he also failed, as we have seen.

The largest single section within the half of his manifesto devoted to science was given over to the virtues of the film in teaching surgical technique, concentrating on films produced by Louis Doyen, whose distribution rights Urban had acquired. But, of the audience for these films, he asserted 'IMPORTANT: Under no circumstances can Surgical or Medical Film Series be supplied for exhibition except to Medical Colleges, Hospitals and Cognate Institutions' (Urban 1909: 83). Again, he was making a claim for the scientific potential – as yet unfulfilled – of the medium. He said in a special supplement, 'our chief desire is that these films should not be inspected and criticised solely as demonstrations of surgical skill, but as showing the possibilities ... of this method of disseminating knowledge of surgical procedure' (1906: 3). In bacteriology, he made great play of the fact that his company's innovations permitted the display of living organisms. Urban's argument for films of bacteria and insects was instrumentalist; he stressed the utility of accurate knowledge of these life forms for doctors, the Board of Agriculture,

Medical Officers of Health, Sanitary Inspectors, sewage disposal committees, farmers, dairymen and brewers. 'The farmer, with the knowledge gained by the cinematographic study of his subject, is better able to cope with the depredations of the numerous insect foes which all too readily devour his small profits' (1907: 43–4). In zoology, films could reveal animal movements, especially of those in the wild, and their facial expressions. Urban argued that it had become 'possible to place before Natural History classes "living pictures" of animal and insect life which … impress the minds of students' (1907: 47). A similar case was made for botany: 'The Cinematograph has led to the observation and chronicling of what may be termed subsidiary phenomena which were heretofore unnoticed, and it is most useful in physiological botany, as showing the movement of plants … Cinematography also teaches that the more we study by means of modern methods, the more the student is struck by the almost human-like instincts of the plants under observation' (1907: 50). The evidence of Urban's publications is that, across the range of science, technology and medicine, the types of science that mainly concerned him were natural history, biology and medicine. Use of the cinematograph in physics is afforded only one paragraph in the manifesto, quoting films on crystal growth, high frequency voltage discharges and the formation of smoke vortices (1907: 45).

Urban's catalogues of films for public exhibition followed the same pattern, with a similar dominance in the scientific films of biological and natural historical subjects.[8] For example, 22 of the 201 pages in the 1906 Catalogue were devoted to 'Marvellous Natural History Subjects', including series on *Empire of the Ants, Quaint Denizens of the Insect World, The Life of the Bee* and *Noah's Ark: Life in the Animal Kingdom,* 'by courtesy of Lord Strathcona and the London Zoological Society' (1906: 138). Each of these sequences was composed of a large number of short – often one-minute – films. The catalogue description of one example, *Chimpanzees at Play,* indicates the terms in which they were expected to be seen: 'These great anthropoid Apes are intensely interesting creatures. Their intelligence is of a high order, and it is very pretty and laughable to watch them play' (1906: 143). Anthropomorphism is not far below the surface in a film named *Baby Chimpanzee and Keeper*: 'The facial expression of the Chimpanzee is most wonderfully human. It is perfectly easy to see by the emotion on the face what the little creature is thinking about' (ibid.).

The press followed Urban's lead in perceiving these films as scientific. The different *ways* in which they may be seen as such are worth considering, not because we will thereby achieve a sense of the essence of the scientific film, but because by exploring the issues we may better understand the ambiguous territory that these, and other films representing science, occupied.

Urban played to the aesthetic of astonishment with his natural historical actualities, and especially with the microcinematographic films (see Gaycken 2003: 36). This is clearly present in how the *The Unseen World* was advertised and perceived by the press. In the *Amateur Photographer*'s 'Notes and Comments', for example, the micrographic films' degree of magnification was stressed: 'An enormous, almost an inconceivable, magnification is necessary to render these minute objects visible to the naked eye … When this photograph is placed in the lantern and projected … the magnification is increased to thirty-eight million diameters' (Anon. 1903b: 1).

Although Urban made claims for the value of his films in science and surgery, there is little sign that the scientific press at the time saw his activities as being of direct concern to them; the scientific community's leading general journal *Nature* represented the show at the Alhambra as being for the public:

> We are glad … that science is being introduced – even in the form of amusement – to those who, in ordinary circumstances, take no interest in scientific matters … 'The music halls are,' says our correspondent, 'being increasingly used for good music; why not for good science? … Those interested in science need not spend the evening there; they could go to see just what concerned them.' (Anon. 1903c: 396)

If we assume the reviews that Urban quoted in *The Cinematograph in Science* to be indicative of how he wished the films to be seen, it is striking that only one other mainstream academic title, the *Lancet*, was quoted. Here the review was limited to clinical films of epileptic seizures and other neurological conditions, not the films that the public saw. Amongst reviews from non-scientific journals, there were several allusions to science: the *Morning Post*, for example, ventured that films 'must be of value to science, not only as an automatic and unerring record of experiments, but as a potent aid to the dissemination of knowledge' (quoted in Urban 1907: 56). The *Sunday Sun* held that 'the cinematograph promises to become [the average Londoner's] most valuable mentor in applied science' (quoted in Urban 1907: 54). Scientific approbation also featured; the *Court Journal* reported that 'the fellows of the Zoological Society last week viewed with solemn delight the micro-cinematographic reproduction of the protoplasms in the cells of the Canadian pond weed' (quoted in Urban 1907: 55). The generality of the quoted reports from lay journals, however, centred on a small number of themes; they were 'instructive without being dull' (quoted in Urban 1907: 53), created with difficulty and the aid of specialised apparatus, and children and adults could learn with ease from viewing the films.

Meanwhile, the audiences that they actually did continue to attract were in the theatres and music halls. The films would have appeared to the majority of those who went to see them at the time as a new kind of novelty, a new amazement to compare with the bicycling cockatoo. But, especially if they were readers of the press accounts that Urban had arranged, they are likely also to have identified them as scientific (see Gaycken 2005: 32–8).

Science in a lay tradition

One approach to the scientific status of *The Unseen World* and *Secrets of Nature* films is suggested by the 'Notes and Comments' column of the *Amateur Photographer*, which stated that the Urban-Duncan films made it 'possible to watch upon the screen the actual processes of protoplasmic life, and show a large audience *what has hitherto only been observable by scientists*' (Anon. 1903b: 41; emphasis added). There was a precise parallel statement on surgery in Urban's manifesto: 'The Cinematograph now renders it possible to reproduce endlessly, under circumstances which permit most close and leisurely study, scenes which formerly could only be witnessed in the operating theatres of our hospitals' (1907: 27). We might argue that Urban could claim the status of science partially because *presenting the subject* of science – that is, the natural world – has often imperceptibly been conflated with the *doing* of science, whilst disregarding – or being denied access to – its methods, institutions and theoretical structures. But the simpler explanation is that, by virtue of the recruitment of Duncan and Smith, Urban was creating an authentic popular science independent of élite biology. For Duncan and Smith, I suggest, the making of these films represented an extension of technique. The *showing* of the films to the public can be seen as a popular variant on giving a paper illustrated with lantern slides to the Quekett Club. This was amateur or popular science made accessible to significantly greater numbers than a Friday evening Ordinary Meeting at the Club. The significant difference was that film audiences, unlike the congenial fellows of the amateur club, could not be expected to be collectors or microscopists. For Urban, it served commercial interests to present himself as sober and serious, because this provided the opportunity of opening up a new market in Britain. No better badge of seriousness presented itself than science. In coming to this conclusion about the relation of these films to élite science, we have reached the same judgement as that which the *Documentary Newsletter* made about Smith's efforts in 1941: 'although correct in detail, they are not made by experts or scientists, but by ordinary people for ordinary people and are therefore not dull, opinionated, high-brow or condescending' (quoted in Burt 2002: 129).

It is important to stress that placing the science and films of Duncan and Smith into the categories of 'amateur' or 'lay' – and their style of invention into 'tinkering' – is not to make a value judgement about their status. Rather, it is to locate them on the rapidly changing map of science in British public culture at the beginning of the twentieth century. This was a period in which science and invention were being professionalised, and these individuals simply belonged to older, not lesser, lay traditions. In fact, we may invert the assumption and celebrate these films and their successors as true examples of lay expertise in science. In subsequent chapters we shall, by comparison, see how professionalised scientists and technological concerns used the cinema.[9]

Popular science films after Urban

Urban was not alone in his linking of science and cinema in this period. And he was either typical of – or perhaps highly influential over – early discussions of the role of the cinema in relation to science, as may be seen by the example of Leonard Donaldson, who made great claims for 'the achievements and possibilities of cinematography as an aid to scientific research', as the subtitle of his book *The Cinematograph and Natural Science* (1912) put it. The tone of this compendium is close to that in Urban's manifesto, which he quotes at some length in the relevant chapters. His preface asserts:

> The 'motion picture' has been enlisted into the service of science to unravel the manifold mysteries of life. In its application to microscopical investigation, and to the study of physical phenomena, it has perhaps proved of the greatest value. It has been found ... sufficiently wonderful as a potent aid to the dissemination of knowledge to excite the marvel of the world. (1912: 7–8)

In addition to the natural history films distributed by Urban and Pathé, he listed film of a volcanic eruption, Professor Störnier's filming of the Northern Lights and use of the cinematograph and phonograph on the Torres Straits expedition. There are descriptions of the filming of wildlife that would not be out of place in a boys' adventure book: 'for days Mr [Cherry] Kearton lurked hidden in the rushes by the banks of the Tana River, watching the movements of the hippopotami and crocodiles' (1912: 45). Kearton and Oliver Pike were amongst the small number of filmmakers who, in addition to Duncan and Smith, brought observational films of animals to public screens in Britain (see McKernan 2005: 355; Bottomore 2005: 521). France and North America also had lively cinematic cultures in which scientific films broadly similar to the Urban type also featured (see Gaycken 2005).

Secrets of Nature

The most famous, long-term and successful of the nature film series was, however, *Secrets of Nature*, produced by Harry Bruce Woolfe's company, British Instructional Films, between 1922 and 1933. The series continued under the name *Secrets of Life* after this company closed and Woolfe, taking key staff with him, moved to Gaumont-British Instructional (G-B I) (see Low 1979a: 24). The films produced in these two series continued the two dominant traditions established in the Urban-Duncan Micro-Bioscope films, of microcinematography and of animal behaviour. Percy Smith was amongst the nature filmmakers that Woolfe employed throughout, and it is with Smith that *Secrets of Nature* is most closely associated. Mary Field acted as producer for the series from 1927 and became its editor from 1929. She and H. R. Hewer undertook directing duties on the films, freeing Smith to concentrate on the intricate business of creating and running the gimcrack combinations of equipment that produced the speeded-up, slowed-down, close-up and micro-cinematographic sequences that were the staples of these films. Initially silent, with caption boards explaining the visual sequences, these films also appeared from the early 1930s with soundtracks.[10] In this way, films shown

Fig 1.3 *Secrets of Nature*: Frank Percy Smith's specialised filming technique reveals the life history of the frog (1929)

as educational devices in the classroom were made into entertainment films in public cinemas when accompanied by dance band scores and commentary by speakers including the Gaumont-British news commentator E. V. H. Emmett. Some examples will give an impression of their approach and style.

Directed by Mary Field, *Magic Myxies* (1931) showcases Percy Smith's time-lapse microcinematography of the growth and movement of a variety of slime mould, *myxomycetes*. Like many in the series, it has a dance band style score by Jack Beaver and a commentary with a slightly facetious tone. It starts:

> Funny little things like tiny tentacles are sometimes to be found on dead wood or on decaying leaves. These little growths are called 'myxies'. Part of their life they are vegetables and part of their life they are animals. And, probably, they would be minerals too if they could ... To get some idea of how tiny they are, look at this picture. The thing like a barge pole in the front is a human hair. And yet, even at this magnification, the myxies are those tiny specks moving about in the background, almost too small to see.

The film continues to discuss the myxies' feeding habits, how in the animal (wet) state they split in two, combine, first in pairs and then in clumps. The tone is whimsical: 'Only pairs are eligible for joining a group. So, if a myxie has been so bad-tempered that it has failed to find a partner, it is not allowed to become one of a party, but is eaten up; this is a far greater encouragement to matrimony than any tax on bachelors.' The vegetable (dry) phase follows, with a description of the movement of a colony of cells, using metaphors of 'flow'. Its response to light is described, along with its feeding habits and its response to a drop of arsenic and another of Epsom salts. The organism's suspended animation in dry seasons is described, as is its response to freezing temperatures. The commentary concludes by describing the growth and dispersal of the myxies' 'fruits'.

Another in the 1931 batch directed by Field was *World in a Wine Glass*, featuring Smith's microcinematographic work and animated diagrams. This is notable for opening its commentary with an appeal to the imagined amateur naturalist in the audience:

> If you suddenly think you would like to have an aquarium, the cheapest way to get one is to fill a glass with water and then put in a wisp of hay. In a few days, if you look through a microscope, you will find your aquarium in full swing. These creatures are called infusoria ... You get an idea of their size when you realise that this group are doing their health exercises within the eye of a very fine needle.

The subject is typical of Smith's aesthetic 'that played on the surprise of the ordinary and reflected the local accessibility of the natural world' (Burt 2002: 129). The film proceeds to describe film sequences of the remainder of the inhabitants of the glass in similarly whimsical and anthropomorphic terms. Once again, the movement and feeding habits of the creatures are described.

The animal behaviour strand is exemplified here by *Ravenous Roger*, directed in 1935 by Mary Field, photographed by Charles Head, with music again by Beaver, and – in the theatrical release version – a commentary by Emmett. (The silent schools version was released under the title *Roger the Raven.*) The film tells the life cycle of ravens on the cliffs of Pembrokeshire. Emmett's commentary describes in facetious tone what is seen on the screen; its role is to entertain, as there is little scientific content on the soundtrack. There is a stream of anthropomorphic jokes (which Emmett had the reputation of making up as he went along). In this case, it does not require the close discourse analysis of a Donna Haraway to reveal the human gender roles projected onto the animal kingdom (see Haraway 1984). The implied scenario is that a group of men on the cliffs are seeking to steal baby ravens. Emmett comments on one of the men:

> by the way his ears are wiggling, it looks like he's seen a blonde going for a bathe ... oh no, it's mother raven on her nest of chicks; that won't make him fall over the edge. The raven at the top is her mate, who went out in search of food. But, as he's rather a long time, mother raven thinks she'd better join him. Wives are like that ... She isn't at all sure that she isn't losing her grip on the old man. And anyway, you can't be too careful; no girl can keep her glamour whilst she's sitting on eggs.

One of the five eggs is identified as containing 'Roger, our hero'. Mother brings food to the raven chicks. 'Father is helping too, but you know what men are. It'd really be much better if he kept out of the way. Why don't you go and cut the grass?' There is more detail about the diet and feeding, interspersed with Emmett voicing the baby chicks: 'More!' There is a description of some of the other marine wildlife including guillemots. The commentary returns to the (older) Roger, learning to fly, 'out for a pilot's certificate, like father'. The film's last two minutes are occupied with the men trying to steal ravens, with Roger being put in a canvas bag before escaping and the thief becoming 'the laughing stock at the Pig and Whistle'.

The films were popular enough to generate a successful sideline in books (Field & Smith 1934; Durden *et al.* 1941; Field *et al.* 1952). These adopted a similar tone to the films; their aim was 'to blend together into one narrative three sharply contrasting points of view; these are the critical accuracy of the scientist, the exuberant enthusiasm of the naturalist, and the anthropomorphic ideas of the layman who strives to translate the doings of an insect in terms of human endeavour' (Durden *et al.* 1941: 9). Setting themselves apart from élite scientific discourse, they aimed for a 'simple and readable' account of 'a subject which, when expounded by those learned in science, is apt to be not only incomprehensible but intolerably dull' (ibid.). This point raises the issue of the comprehensibility of science for lay viewers of scientific films. Here the issue is sidestepped by avoidance of technical language. In the versions of films made for the general public, as we have seen, it was evaded by the use of homely anthropomorphic language and whimsical wise-cracking. But contemporary critics did not reject the films for not being scientific. Instead they were almost universally praised; John Maddison for example alluded to Smith's 'brilliant and sustained experiment' (1948: 118).

Conclusion

The growth and specialisation of science and technology provide a context for the story of science filmmaking throughout the twentieth century. Although cinema did not entirely usurp the public culture of science, it did become a universal site to experience the spectacular, and within that, science as non-fiction had a significant part to play as much as fiction. But this was a sector of culture that had its own structure and which was, at this stage, scarcely connected to professional science. This is clear in the absence from public screens of the major developments in the physical and biological sciences, including relativity, quantum mechanics and Mendelian genetics. We must recognise that *The Unseen World* or the *Secrets of Nature* series were only the first mode of filmmaking to represent science, technology or medicine.

2 TECHNOLOGY, SCIENCE AND MODERNITY: THE WORLD OF DOCUMENTARY

[A] complex set of attitudes and practices of the progressive intelligentsia in the interwar period within which scientific and technological progress (modernism in all its varieties) and the new forms of communication ... were seen as combining a new educative and democratic potential. (Scannell 1980: 2)

Non-fiction films and the human world 1900–39

In early July 1933, following dinner, the delegates to the World Economic Conference sat down to a performance of films selected to showcase British cinematic achievement. At the culmination of the first half came the first showing of the director Paul Rotha's film, *Contact*, sponsored by Imperial Airways (see Rotha 1973: 91–2). The Prime Minister, Cabinet members and assorted dignitaries were presented with a film, accompanied by music and some engine noise, with bold titles in the Russian style: 'For a Long Time ... Sea ... Road ... Rail ... Now Air ... Aircraft are Planned ... The Raw Materials ... Constructed ... While ... Bit by bit the engines are built ... tested ...' intercut with a montage of men at work, culminating in shots of an aeroplane at Croydon aerodrome. The film continued with an extended impressionistic travelogue, a journey through the Middle East and back through Africa to London, interspersed with engine shots and views of planes from the ground. The audience had witnessed a paean to the power of the new technology of flight, a highly aestheticised account of the power of humans to transform their worlds by the development of new machines. But how novel or typical were such films? And how did these 'documentaries', as they were called by their champions, differ from other films representing science and technology?

Evidently, nature was not the sole subject of non-fiction films in this period. In fact, human activity had also been a highly significant subject since cinema's first decade. This started from the time of Charles Urban and increased throughout the interwar period. Notable subjects were the emblematic technologies and industries of the era – communications, coal, steel, carmaking, shipbuilding; even the countryside was often shown as transformed by scientific practice and new technologies. Well before the late 1920s, filmmakers were producing travel, exploration, advertising, health propaganda,[1] war reconstruction, 'interest', instructional films and newsreels, many of which represented technical subjects (see Low 1971: 285-97). Charles Urban's catalogues list substantial sections on 'industrial', 'naval' and 'railway' films. Many of the more than seventy 'industrial films' in the 1909 catalogue, such as *Trapping Salmon on the Fraser River*, overlap with travelogues. Others, such as *The First Electric Train out of Paddington Station* (before 1909) or *A Bird's Eye View of Paris* (1910), record technological wonders. *The Blackpool Aviators' Meeting* (1909) records a few details of an event, whilst *The Motor Climbing Contest at Crystal Palace* (1904) is simply a sequence of six shots showing cars driving up steps. *A Visit to the Steel Works* (before 1909), 450 feet in total, comprised 23 short films from '1 – unloading ore at the iron works' to '23 – making a chain cable for the British Navy'. The evidence of the shot list is of a sequential narrative covering all the significant operations, a literal account of the process (see Urban 1909: 61–2). Urban commented 'in picture form, the veriest dunce is compelled to evince an interest in matters of everyday life' by seeing the processes involved in the production of a newspaper or the construction of a railway (1907: 20–1). His Kineto Company, which he established in 1910 to concentrate on scientific films and travelogues, also anticipated the concerns of documentarists by making a film in 1910 called *A Day in the Life of a Coal Miner*. This, and films by competitors, such as George Cricks and Henry Sharp who made *A Visit to Peek Frean and Co's Biscuit Works* in 1906, had developed from the simple sets of shots of the earlier films. The mining film has a sequence of shots of activities at the mine, featuring women and men workers, within a 'day in the life' structure framed by a miner's departure from and return to his home; a final shot shows a bourgeois family enjoying the fruits of his labour. The Peek Frean film also has a sequenced narrative, starting with the delivery of ingredients and tracing the manufacturing process all the way to the vans carrying the finished products from the factory.

In 1924 Pathé made *The Imperial Airway: The Work of the British Airways*, whose titles refer to it as an 'Air Ministry official film', an early example of the government funding that was so distinctive a feature of the later documentary units. This example indicates the style and concerns of 'interest' pictures at this date. It starts with a long caption which first asserts that air transport satisfies the need for speed and safety, and then introduces the first shot – of passengers and luggage being loaded into a car at the Hotel

Victoria. We then see a thirty-second static shot of this operation. This turns out to be a framing device, as the next caption asserts that 'meanwhile, the aerodrome at Croydon presents a very busy scene'. Hereafter, the film follows a 'day in the life' structure. The work of the aerodrome is shown in a series of long shots (both in camera view and in duration), which are mainly static, except where movement within frame (such as take-off) demands some panning. Cutting between scenes and captions is businesslike; shots are held long enough to convey their point, and some transitions between shots are enlivened with irising-in or -out. The expertise of ground engineers in checking each plane before flight and certificating it as fit to fly is asserted. We see the work of the Meteorological Office in checking weather conditions and wind-speed (by tracking a balloon using a theodolite) and in the posting of weather reports on a large notice board for pilots to inspect. At this point, we see the passengers from the first scene arrive and go through the business of baggage loading and embarkation before a plane is seen taking off. The comfort of the craft is asserted and we are shown the passengers seated within. There are aerial shots of the aerodrome and of 'the garden of England'. An animated map with captions shows how much faster air travel to European destinations is than the alternative of sea and rail. The use of radio for communication with the ground, and by the use of bearings to calculate an aircraft's specific location, is explained. Finally, we see Croydon aerodrome in the dark, illuminated via the use of firework rockets to aid night landings. One such descent is shown and a final caption reads 'nothing is omitted to ensure a perfect and safe landing'. Here is a film, securely in the idiom of the interest picture, but which has gone well beyond the early Charles Urban films, especially in its story structure. In its representation of science and technology, it shows not just the aeroplanes themselves – still something of a wonder of the age in 1924 – but also the expertise of ground engineers, meteorologists, pilots and those who plot the aeroplanes' course as integral to providing a swift and safe mode of transport.

Non-documentary 'interest', promotional and instructional films continued to be produced in the 1930s, in parallel with documentaries. The films sponsored by the Central Electricity Board (CEB) and Electrical Development Association (EDA) provide an example. They employed Bruce Woolfe's company, amongst others, to make 13 films from the late 1920s with titles including *Electricity* (1927) and *Rural Electrification* (1936); these were 'silent films that were in fact little more than lantern lectures, with long, informative titles replacing the lecture' (Low 1979b: 141). Populist EDA films included *Plenty of Time for Play*, *The Wizard in the Wall*, and *Well, I Never!*, all directed by Alexandre Esway for fellow Hungarian Alexander Korda's London Film Productions (Anon. 1934a). The first of these stands out for its promotion

Fig 2.1 ***Plenty of Time for Play*****: released in 1935, this film used a futuristic vision of life in 1955 to promote electricity**

of technological modernity. It was set in the future electrified world of 1955, in which the need for work was portrayed as significantly curtailed and domestic drudgery almost eliminated. Its fictional scenario featured Geoffrey, a modern man, who shares his office with a conservative character, Gribble, who is writing a book on life in the 1930s. The world of 1955 features videophones, travel by private autogyro, machine cookery and a giant television screen. Towards its end, the film is interrupted; it turns out that we have been watching a film within a film. The actor playing Gribble drives home the message that electricity can already in 1935 make housework cheaper, quicker and more efficient.

In the production of factual films on the human world, Gaumont-British Instructional (G-B I) were as active as they were in natural history. Mary Field, for example, made *The Farm Factory* for G-B I in 1935, which represents agriculture as an industrial process. But the evidence is that this film, and others including J. B. Holmes' *The Mine* (1936), were indeed instructional, aimed not at the general adult public, but at schools. Documentaries, to which we now turn, did aim specifically at adult audiences.

The documentary mode

From the release in 1929 of *Drifters*, John Grierson launched 'documentary' as a form of non-fiction filmmaking that he and its other champions represented as new and different from the older types of factual films. From this beginning, a sense of documentary has been derived that has been shared by filmmakers and both historical and theoretical writers. Often, in the way that it is discussed, there is more than a suspicion of Platonic idealism, of 'the documentary' as a category beyond individual films. This idealist tendency deflects attention from the contingent historical realities that created the films and leaves closed the black box that a truly historical account must open. It is not difficult to see how such a Platonic notion of documentary came to dominate the discussion of factual films. British documentarists, who were seeking to differentiate their practice from that of other filmmakers, invested significant energy in representing it as an entity with specifically new characteristics. It is arguable that this came intuitively to John Grierson, whose training in idealist philosophy, which has been so well illuminated by Ian Aitken (1990), may have predisposed him to think of documentary as an ideal beyond the worldly approximations he and others produced as actual films. It is necessary to acknowledge that this idealism was part of the documentary rhetoric, but also to avoid falling into believing it to be true: documentary films are simply the products of their historical circumstances. This is not to deny that documentarists made films that differed from others; it is to assert that the differences are concrete and historical, not abstract and universal.

The most significant consequence of the idealist account is a deflection away from the corpus of films themselves. For the current work, the major deficiency of the existing historiography is insufficient analysis and discussion of the fact that British documentaries consistently represented machines, the relationships of people to machines, rationality and rationalising responses to health issues.[2] Particular technologies were widely considered to be emblematic of modernity in this period. The editors of the collection *Cinema and the Invention of Modern Life* have suggested that 'Modernity … has been familiarly grasped through the story of a few talismanic inventions: the telegraph and telephone, railroad and automobile, photograph and cinema' (Charney & Schwartz 1995: 1). We might, deferring to Lenin and Le Corbusier, add heavy electrical technology, broadcasting, ocean liners and steel-framed buildings to this list. Most of these technologies were featured as the subjects of documentaries; for example, aviation in *Contact* (1933) and *Airport* (1935), broadcasting in *BBC: The Voice of Britain* (1935), railways in *Night Mail* (1936) and transport of all kinds and hydro-electric technology in *We Live in Two Worlds* (1937). Pre-eminently, of course, the emphasis in the General Post Office (GPO) Film Unit films was on telephones and telegraphs.

Part of the reason for this lack of coverage of the films has been an implicit conflict at the heart of discussions of British documentary between definitions that stress people and those that emphasise the theory and practice of filmmaking. Following the original apologetics of Grierson, it has been common to use what might be called a 'genealogical' definition of British documentary, that is one that stresses the person of John Grierson and the individuals and organisations with which he was associated. Anyone who was at the Empire Marketing Board (EMB) or GPO units, that is a member of Grierson's 'family', was by this definition a documentarist. 'Genealogical' accounts marginalise documentarists such as J. B. Holmes, who moved between instructional films and documentaries, and Paul Rotha, who worked outside the government units. Emphasis on government units has also accentuated the disproportionate coverage given to films made under Grierson's tutelage, which has often been limited to a small canon of 'classic' documentaries, focusing for example on Grierson's own *Drifters* (1929), Basil Wright's and Harry Watt's *Night Mail* and a selection of works by Humphrey Jennings. Despite the broader view of Brian Winston (1995), very few of the GPO Film Unit's approximately seventy films, let alone the more than 250 documentaries from the 1930s listed by Low (1979a), have so far received extensive scrutiny or discussion. The greater the emphasis on association with Grierson, the less coverage will other films and filmmakers receive, however deserving they are. But, if permitted into consideration, the wide range of films within the GPO canon make our argument for us: their diversity equals that shown by all other non-fiction films taken together. Within the output are Surrealist whimsy (*Pett and Pott*, 1934), instructional films (*How the Dial Works*, 1936), formal experiments (*Rainbow Dance*, 1936), documentary drama (*Savings of Bill Blewitt*, 1937), English ethnography (*Spare Time*, 1937), recruiting films (*Job in a Million*) and magisterial portraits (*Health for the Nation*, 1939). Merely selecting different films for analysis – as this book does – produces a markedly different 'documentary story'.

'Genealogical' definitions within the documentarists' writings sit very uncomfortably with those that stress the formal qualities of documentaries. These were the filmmaking techniques that followed from 'the creative treatment of actuality', Grierson's coinage for the formal distinctiveness of documentaries. If this phrase is taken to allude to documentarists' higher level of cinematic literacy than their competitors, including an engagement with the Russian style, then many simply do not belong. Viewing the films themselves, and especially those that are less well known, reveals many that – even to the trained eye – could just as easily have been made, for example, by G-B I as classroom films. One of many examples is Marion Grierson's *For All Eternity* (1935) – a historical survey of British churches – which employs a simple historical sequence, unpretentious and artless framing of shots and a plain, literal, editing style undistinguishable from that used by any of the small commercial film companies with fewer aesthetic aspirations. So, emphasis on the documentary 'family' creates a field of filmmaking practice that is so wide as to overlap with the non-documentaries we have already discussed. Equally, a focus on those films that exhibit particular formal qualities will tend to exclude many films universally accepted to be documentaries.

Both the genealogical and the formalist definitions characteristic of British documentary studies divert attention from what we may call an iconographic approach to the subject, one that privileges the ways that documentary filmmakers represented the world to audiences. As is demonstrated in the following pages, this latter approach can produce a significantly different picture of documentary by revealing the centrality of its emphasis on science and technology. One result of this shift of focus is to move away from considering the task of the historian-to-be to write *the* history of *the* documentary film movement. Instead, it implies that documentaries should be components of many histories and these should focus not on 'the movement' but on the world that documentarists represented.

The idealist, genealogical and formalist accounts have been influential in large part because of the highly rhetorical writings of Grierson and Rotha in particular. Behind these apologetics lay concrete financial issues. The documentarists' economic standing was often precarious, both at the margins of government as policies fluctuated and in the commercial market, which contained many filmmaking companies effectively competing for the same contracts (Rachael Low says there were as many as sixty in the 1930s; see 1979b: 137). Grierson and Rotha were highly-skilled prose stylists producing vividly rhetorical language that was aimed to recruit support for their enterprise. They asserted the distinctiveness of documentary through a mixture of aesthetic discrimination and some excusable ignorance of other work.[3] They invented a tradition for their oeuvre, not least for the sound commercial reason of 'product differentiation'. And their writings set the terms for how documentary came to be seen not only by supporters, but also by detractors. As Rotha's books in particular became standard works, some older types of factual films, including those discussed above, became historically marginalised.[4] Also, later self-defined schools of cinema – such as the 1950s Free Cinema – contented themselves with asserting the novelty of their own practice, rather than rediscovering what documentarists had elided.

There were most often overarching factors governing the choice of subjects. The films made by the GPO Film Unit from 1933 were marked by a new concern with 'representations of modern industry, technology and mass communications' compared with 'the preoccupation with rural and regional experience in … earlier [documentary] films' (Aitken 1998: 12). But, if the iconography of their work at the GPO was more obviously technological, that was substantially because of the concerns of the GPO, especially in telephony. Equally, because of the EMB Unit's requirement to represent Empire produce, there had been an inevitable focus on agriculture, the countryside and the natural. With the representation of health or medical subjects, the same factor was at play. Rotha said 'we may assume … that documentary determines the approach to the subject but not necessarily the subject itself. Further, that this approach is defined by the aims behind production, by the director's intentions and by the forces making production a possibility' (1936a: 134). Technological subjects were also included in the interpretation of other themes; for instance *Health for the Nation* (1939), in making an argument about public health, devotes a significant proportion of its length to showing industrial technologies such as coal mining and railways.

To question the genealogical account is not to dismiss the importance of social historical explanations for the films' subjects and styles. Both finance and the cultural contexts that influenced the films' treatments of their subjects rested on relationships between people. Filmmakers required funding, access to subjects and to people who would advise and appear within films. Accordingly, it is clear that the interests of filmmakers alone are insufficient to explain the fact that individual films were made. In this period, documentary was not so well established that scientists, doctors and technologists often actively sought opportunities to become involved in films (the major exception to this is the subject of the next chapter). Members of these professional groups became involved when they were drawn into productions as part of the extended networks necessary to bring films to fruition. This might be as employees of sponsoring organisations, as individuals invited to take part and appear on screen, or as suppliers of opinions, expertise or moral and political support to filmmakers. And so groups of people with disparate politics – as well as other beliefs – often agreed to collaborate in the production of a film for the sake of its perceived persuasive power. The authority of participants helped gain support, and funded films helped provide a platform for the expression of interests by participants. Reciprocally, filmmakers had to appeal to the interests of heterogeneous organisations and individuals. These might include corporations, government departments, scientists, politicians and many others. Support amongst politicians for public relations, and for the documentarists, was not tied by party affiliation. Kenneth Lindsay, a key figure in Political and Economic Planning who was a National Labour MP, was a close associate of the documentarists. The Conservative MP Walter Elliot was also a long-term supporter.[5]

Asserted distinctiveness

Paul Rotha's viewpoint provides a valuable guide to our subject, engaged as he was in the making of significant films that fused a particular view of science and rationality with a deep engagement in filmmaking technique. With Grierson he was one of the most prominent figures in establishing documentary filmmaking in Britain. He had trained at the Slade School of Art, worked briefly as an art director for British International Pictures, before writing at the age of 22 *The Film Till Now*, an influential textbook on the history of films, published in 1930 (see LeMahieu 1988: 218–22). In 1930, after six months working with Grierson at the EMB Unit, he went freelance. In 1932–33, with support from Shell Petroleum and Imperial Airways, he made *Contact*, his first substantial documentary. Over the next quarter century, he made many significant films on scientific, technological and public health subjects. But his influence was as much via his prolific written work as through his films.[6] His *Documentary Film*, published in 1936, was the first book-length justification of the genre in Britain, representing documentary as a distinctive social and political mode, literate in cinematic technique.

Rotha distinguished documentary firstly from fiction films, and secondly from other types of non-fiction. For him, documentary was not simply the British school of Grierson and the government units, but a non-Hollywood tradition in which film technique was applied to the interpretation of social themes. His version of documentary 'had its real beginnings with Flaherty's *Nanook* in America (1920), Dziga Vertov's experiments ... Cavalcanti's *Rien que les Heures* ... Ruttman's *Berlin* ... and Grierson's *Drifters*' (Rotha 1936a: 77). Using a rejection of Americanisation typical of British intellectuals at the time, he presented documentary as a serious-minded antidote to the frivolity of Hollywood (see LeMahieu 1988: 118–21).

> The story-film has followed closely in the theatrical tradition for its subject-matter; converting, as time went on, stage forms into film forms, stage acting into film acting. The opposite group of thought ... proceeds from the belief that nothing photographed, or recorded on to celluloid, has meaning until it comes to the cutting-bench; that the primary task of film creation lies in the physical and mental stimuli which can be produced by the factor of editing. (Rotha 1936a: 76–7)

Fiction and non-fiction were expected to engage audiences in entirely different modes:

> [Documentary demands] from an audience an attention quite different from that of a fictional story. In the latter, the reaction of the spectator lies in the projection of his or her character and personality onto those of the actors playing in the story and the ultimate result of a series of fictional complications ... [whereas] in watching documentary, the audience is continually noting distinctions and analysing situations and probing the 'why' and the 'wherefore'. (Rotha 1936a: 141–3)[7]

Making his distinction of documentary films from existing non-fiction, Rotha was determined to assert the superiority of documentary: 'the step that exists between this type of general "interest" picture and the higher aims of the documentary method is wider than is usually imagined' (1936a: 75). He commented dismissively that, where industrial propaganda films were being made by commercial film companies, 'they have most often been regarded by the Trade more as an easy means of making profit than as an opportunity to develop a new branch of cinema' (1936a: 52).

Social actuality

Rotha linked 'the actual' and 'the social' in two specific ways. The first kind of social actuality formed the grounds for Grierson's philosophy. As Ian Aitken has lucidly conveyed, Grierson advocated a type of philosophical realism, derived from his university education in idealist philosophy, which went beyond the world of appearances to a deeper underlying reality, in a 'prioritisation of essence over experience' (1990: 110):

> In documentary we deal with the actual, and in that sense with the real. But the *really real*, if I may use that phrase, is something deeper than that. The only reality which counts in the end is the interpretation which is profound. (Grierson quoted in Aitken 1990: 109)

Aitken gives an example of the intercutting of shots of a herring trawler at sea with shots of the market place in Grierson's own film *Drifters*, forming a sequence which is designed to show their interdependence and the 'interrelation of social practices within a social reality' (1990: 110). He also, Aitken argues, 'wished to express the reality of the way in which "simple heroic labour" was transformed into a market commodity' (ibid.). Rotha gestured to similar concerns when he wrote that 'the world of documentary is a world of commerce and industry and agriculture, of public services and communications, of hygiene and housing. It is a world of men and women, at work and leisure; of their responsibilities and commitments to the society in which they live' (1936a: 17). This was the dominant mode of documentary until the mid-1930s. Rotha described this as the 'poetic' style (1952: 110). So, in the first place documentarists filmed actuality and rendered it social by means not only of representing relations between different components of reality, but also, in the showing, represented one part of society to another. In these films, as exemplified by *Industrial Britain* (1933), the documentarists were concerned to show the dignity and craftsmanship of the worker as a means to educate the audience in the nature of modern industrial and corporate society. Here were fishermen, shepherds, mechanical engineers, ship builders and miners.

The expertise of the director was, in Rotha's view, tied to engagement with the analysis of social relationships, not just at an intellectual level:

> Before [the director] can create, before he can become in any way significant in his work, he must be able to understand the social relationships contained in his theme and be dynamic in his social analysis. Not only must he feel his subject and its implications in his head but in his heart. (1936a: 223)

For him it was social *problems* or the relations between social classes, as much as social interactions in the Griersonian sense, that defined his view of actuality. There was an emotional as well as an intellectual response to the Depression, which he shared with many left-wing intellectuals. He wrote to his friend Eric Knight whilst he was shooting *The Face of Britain* (1935):

> On the way here everything was marked by tragedy. All through the industrial Midlands and Lancashire, the terrible slums of Glasgow and the Clyde Valley, you could see the scarred mess that greedy men have made of this lovely country. And so much of it is now disused. Great slabs of countryside littered with rusting factories and crumbling chimneys, refuse from another age when men trampled upon everything to get rich.[8]

This type of exposure to the impact of the Depression affected the mode of social representations produced by many documentarists from around 1935. In Rotha's work, this was first suggested in the strong implication of the social effects of cyclical unemployment in *Shipyard* (see Rotha 1973: 100–1). More generally, it is seen in the ways that films such as his *The Face of Britain* and Edgar Anstey's *Enough to Eat?* (1936) embody a commitment not only to representing but also to solving social problems such as slum housing, working-class malnutrition and smoke pollution by means of appropriate planning and technical expertise, scientific, social or administrative.[9]

Politics

One of the most pervasive tropes in both contemporary and historical accounts is that of 'progressive politics'. When unquestioningly used by historians – as for example in some 1970s and 1980s writings that sensed a comradeship between the left of the past and the (then) present – this may give rise to a misleadingly whiggish account of the past (see MacPherson 1980). 'Progressive' as deployed by the documentarists denoted a range of categories including various depths of ideological and political engagement with social issues, the wish to make films on particular social themes, the choice of cinematic account and technique, and calls to political action, generally by democratic means. But the core political philosophy was citizenship; Rotha linked this to modernity:

> Civilisation today, in fact, presents a complexity of political and social problems which have to be faced by every thinking person. As soon as politics concern the shape and plan of our civic system, as soon as they concern our very homes and means of livelihood, it rests with the ordinary person to act not merely as a passive voter but as an active member of the State. (1936a: 33–4)

This relied on a sense of the present produced by science and technology:

> The success of science and machine-controlled industry has resulted in an unequal distribution of the amenities of existence under the relations of the present economic system. Side by side with leisure and well-being, there is also unemployment, poverty and wide social unrest. Our essential problem today is to equate the needs of the individual with production, to discuss the most satisfactory economic system and to present the social relationships of mankind in their most logical and modern ordering. (1936a.: 117–18)

Documentary filmmaking became possible in Britain in the interwar period through a historically contingent linking of issues of citizenship with the rise of public relations under the period's particular economic conditions. With the extension of the franchise to women in 1928, the electorate became representative of the British adult population. At the same time, some economists, especially those associated with the New Fabian Research Bureau, were concluding that the world Depression succeeding the Wall Street Crash proved the need for nation states to intervene in national economies (see Durbin 1985: 158); this in its turn implied a new relationship between the citizen and more highly interventive states. Under these circumstances, citizenship discourse became ubiquitous, appealing to individuals across the political spectrum, from the radical left to some Conservatives. In the midst of this, some politicians and officials developed a conviction that it was important to communicate with the electorate. As a result, public relations departments were established in many government departments (see Grant 1994).

The growing respectability of public relations was essential to the establishment and development of documentary (see Swann 1989: 2–4). First at the EMB, then at the GPO, and latterly at other departments with scientific and technological responsibilities, including the Ministry of Health, officials were dedicated to public relations tasks. Under the policy of 'imperial preference', and as a sub-department of the Colonial Office, the EMB's role was to be actively interventive, spending government money on scientific research and promotion of empire products. Under the influence of its secretary Stephen Tallents, use of documentary films was an obvious extension of innovative publicity methods that also included the establishment of nationwide poster hoardings. On the dissolution of the EMB in 1933, Tallents and the documentarists transferred to the GPO, the most technological and commercial of government departments (see Swann 1989: 47–8). Where documentary film was employed in government, there was a marriage of the interests of a particular type of filmmaker, committed to ideas of the film as a tool of state information services, and senior officials' perception of a need to communicate with their public. Rotha wrote that:

> propaganda may become … one of the most important instruments for the building of the State. It is surely only a matter of time before the State will make full and acknowledged use of education, radio, cinema, pulpit and press to ensure public reception of its policies. (1936a: 48)

The documentary ideology involved using films to inform modern citizens about the nature of the world they inhabited, so as to make informed democratic choices, and indirectly to buy into the modernising state and the powers and technologies of the modern world. Some theorists see the expansion of nation states into ever-increasing areas of citizens' lives as a core feature of modernity. Anthony Giddens, for example, notes that 'modernity produces certain distinct social forms, of which the most prominent is the nation state' (1991: 15). This highlights for us the significance of the rational state as a component within many documentaries. Max Weber's famously pessimistic view of modernity described the ways in which modern capitalist societies had increasingly become subject to calculation and rationalisation, a tendency exemplified by the sciences but also by businesses and nation states. In the West, he argued, 'there is a marked tendency to regard rationalisation as desirable in its own right, as an ideal of efficiency which should govern all activities' (see Hughes *et al.* 1995: 96). Within his view of organisations, rationalisation took the form of bureaucracy and there was a tendency for the value of individuals to become lost in vast administrative structures; life became conducted within an 'iron cage' of bureaucracy, regulation and administrative oppression. But the world that Weber feared and distrusted was precisely that which many documentaries represented and celebrated.

For example, Arthur Elton's *Workers and Jobs* (1935), made for the Ministry of Labour, served to promote not just its ostensible subject, the efficiency of the service provided by the labour exchanges, but also the virtues of bureaucratic systems more generally. This is visible in a sequence of shots showing individual workers' cards being removed from their filing drawers. Harry Watt's *Big Money* (1937) brought the world of the GPO Accountant General's department to the attention of the public; *Job in a Million* (1939) promoted junior employment opportunities for would-be telegraph boys; Raymond Spottiswoode's *Banking for Millions* (1935) conveyed the workings of the Post Office Savings Bank; John Monck's *Health for the Nation* (1939) explained the rationality and role of the Ministry of Health. Not only in these films, but also in those discussed in more detail later, Weber's iron cage was unlocked, because rationalisation here was willed and promoted as the way to a better society.

There existed underneath this shared ideology of citizenship a considerable diversity of individual political beliefs and practices. Rotha framed his arguments not within the philosophical idealism of Grierson, but within Marxist dialectical materialism drawing, for example on Raymond Postgate's *Karl Marx* (1933; see Rotha 1936a: 235). The language of the relevant key passages in *Documentary Film* suggests conclusions reached with some difficulty. He quoted Marx from the *Critique of Political Economy*, that 'the mode of production in material life determines the general character of the social, political and spiritual process of life', and proceeded to assert that 'every documentalist [sic] who is producing significant product today is, I suggest, basing his outlook on the belief that it is the material circumstances of civilisation which give rise to and govern the current ideas of society' (1936a: 234). This also shaped Rotha's reasoning for what he perceived to be commercial cinema's failings; using a conventional Marxist base/superstructure metaphor, he argued that:

> we must not forget that it is to the advantage of a dominant class to produce and perfect a form of indirect propaganda for the preservation of its interests. All institutions, whether political, sociological or aesthetic, fundamentally reflect and assist in the maintenance of the predominating influences in control of the productive forces in their particular era. To this the cinema is no exception. (1936a: 45)

Here, Rotha's Marxism, like that of so many of his contemporaries, was naïve and in all likelihood not informed by extensive reading. But that does not detract from the fact that, despite its limited depth, this was the political belief within which he operated.

Dialectical montage

Rotha discussed Hegelian dialectic which, used as historical interpretation following Marx, 'is a way of looking at human history and experience as a perpetual conflict in which a force (*thesis*) collides with a counterforce (*antithesis*) to produce from their collision a wholly new phenomenon (*synthesis*) which is not the sum of the two forces but something greater than and different from them both' (Cook 1981: 170). Rotha was ambivalent about its use at the level of historical understanding; he commented that 'by some modern authorities it is considered an out-of-date method when applied to history' (1936a: 235). But for filmmaking, Rotha was sure: 'dialectical reasoning is concerned with the attitude of mind of the director and not with the subject matter with which he is working' (1936a: 234). In other words, Rotha could choose a dialectical pattern for his material for *cinematic* reasons, just as other socialists did so for *political* reasons. Some of Rotha's difficulty is also evident in a letter to Eric Knight from October 1933:

> Good God, Knight, isn't there some middle way? I know all the Marxist stuff – thesis, antithesis equals synthesis – but somehow I'm not convinced. I know that all art is governed by economics – by the social system – that there's no such bunk as art for art's sake – but somehow I'm not satisfied.[10]

Graham Greene, reviewing *Documentary Film*, commented that 'the first part of Mr Rotha's book, so admirable when it reaches the actual making of documentaries, is rather tiresomely Marxist' (1995: 489). In fact, we may note that both parts of the book are Marxist, a consequence of Rotha's acceptance of the Russian model as he understood it. He believed that the Russian directors and theoreticians Sergei Eisenstein, Alexander Dovzhenko and Vsevolod Pudovkin, had developed a definitive film construction technique (see also LeMahieu 1988: 220). Pudovkin himself had made a scientific record film about Ivan Pavlov's experiments, *Mechanisms of the Brain*, in 1925. For many, Soviet montage theory was considered to be the rigorous and scientific approach to film construction; Ivor Montagu's preface to the English edition of Pudovkin's *Film Technique* compared its impact on film to that of Gregor Mendel's genetics on animal breeding (1933: vii). But Rotha criticised films which did not marry technique to purpose (1936a: 161). This extended to all aspects, including acting; photography: 'technique must come second to content'; and editing: 'the aim of cutting is to stir the emotions of the audience so that it will be receptive of context without the cutting itself becoming prominent' (1936a: 193, 200). After the sections on the visual and aural components of films, he turned to discuss treatment and his recommendation of dialectical montage to structure films, a theory of film construction derived from the Soviet school.

> Eisenstein maintained that in film editing the shot or 'montage cell' is a thesis which when placed into juxtaposition with another shot of opposing visual content – antithesis – produces a synthesis (a synthetic idea or impression) which in turn becomes the thesis of a new dialectic as the montage sequence continues. (Cook 1981: 170)

Dialectical montage was an aspiration for many documentarists; together with their particular models of citizenship and subjects, it provided a formal basis on which to distinguish their films from other cinematic genres. But, in this period, there was little written material available from the Russians on which an aspirant disciple could base their practice. Pudovkin's *Film Technique* had first appeared in an English translation by Ivor Montagu in 1929 and Eisenstein's joint *Statement on Sound* (with Pudovkin and Grigory Alexandrov) had been published in the film art magazine *Close Up* in 1928. It was therefore, for the documentarists, mainly a matter of seeking to reproduce the aesthetic of montage, a process stimulated and fertilised by watching such Soviet films as became available, notably when they were shown at the Film Society, an organisation established in 1925 to screen films not in distribution in Britain. Rotha maintains that Grierson did not fully understand dialectical montage when he made *Drifters*, the film that launched British documentary. This is borne out by Aitken's gestures towards Grierson's 'intuitive understanding' and 'Grierson's frequent references to Eisenstein's use of "symbolic counterpoint"' (1990: 76). 'Counterpoint' was the language of the *Statement on Sound*, which does not allude to dialectic (Eisenstein *et al.* 1994: 234–5). Grierson, having provided the captions for the versions of *Battleship Potemkin* (*Bronenosets Potyomkin*, 1925) and Victor Turin's *Turksib* (1929) seen in Britain, had more close exposure to Russian films than many. But, of all the documentarists, Rotha was probably the most earnest in his attempts to understand and replicate dialectical montage. He was deeply engaged in the technicalities of film construction, to the extent that when the Film Society had a copy of Pudovkin's *A Simple Case* (1932), Rotha borrowed it and analysed in close detail how it achieved its montage effects.[11] He wrote:

> The dialectic as drama is conflict and must dictate the structure of the film. The pattern-of-three arises again and again during production: in the fundamental composition of the film strip (the conflict between frame and frame, shot and shot, etc.), the building up of symphonic movement (comparative rhythms), the imagistic use of sound (two motives expressed si-

multaneously giving rise to a third idea), the structure of sequences and, indeed, quite possibly in the structure of the film as a whole. (1936a: 235)

Others were also successful in adopting dialectical montage. Basil Wright built *Song of Ceylon* (1934) on a dialectical structure (see Rotha 1936a: 230); attempts at emulating the technique and its effects also made a significant contribution to the visual and aural structure of many of the GPO Film Unit films.[12]

In the final part of his discussion of approach and style, Rotha defined two kinds of documentary; 'the descriptive, *reportage* or journalist approach', which he said aimed to be

> an honest effort to report, describe or delineate a series of events, or the nature of a process, or the workings of an organisation on the screen … The less your journalist director sensationalises his material, the better is his purpose served, because by dramatisation he would sacrifice exactness for impressionism. (1936a: 225)

The other method, the 'impressionistic', he stated

> demands just such an exacting understanding of material as the *reportage* method, but it selects only those elements of the subject which are capable of dramatisation. It aims to produce a general emotional effect and not a detailed literary description. It aims to disturb the audience emotionally, to make it feel for itself the social or other references contained in the subject. (1936a: 226)

Rotha stated a preference, committing himself to a dialectical method in impressionistic documentary. His argument for this style was partially founded on the nature of the cinema audience: 'We must make allowance for a tolerant and mildly uninterested audience, upon whom our films must create emotional effect as well as persuade to a certain way of thinking' (1936a: 231–2). Many science films took the alternative route of reportage. But Rotha was resolutely Modernist, devoted to exploiting to the full the potential of the filmic medium.

Industrialisation, planning and modernity: Rotha's *The Face of Britain*

Because Rotha made *The Face of Britain* at the same time as he was writing *Documentary Film* an analysis of this impressionistic documentary can be used to show what the documentary mode in Rotha's hands meant to the representation of technology and associated modern themes.[13] The production documents for the film convey his ambitions:

> The conception of this film is unusual. Documentary in style and approach, it sets out in three acts a dialectic which argues the effect of the mechanisation of industry on our social lives and, after a survey of the chaos produced by unplanned industry, suggests the urgent necessity for a planned and organised future.[14]

The fifth of his documentaries, it was initiated in summer 1934, whilst he was crossing the country to Barrow-in-Furness to record the construction of the liner *Orion* for his film *Shipyard*.[15] It was shot between July and November, and edited between August 1934 and spring 1935.[16] Rotha had to build support for making the film. First he persuaded Bruce Woolfe, managing director of Gaumont-British Instructional, to which he was attached, to agree to the project. Woolfe agreed to pay for footage, on condition that Rotha find sponsorship (see Rotha 1973: 102). Rotha turned to Hugh Quigley, the highly cultured Chief Statistical Officer for the Central Electricity Board (CEB). The benefit was intellectual as well as financial. Rotha recalled that they 'at once saw eye to eye about a film on the possibilities of replanning Britain, both industrial and rural, based on the flexible power of the national grid' (1973: 103). Quigley, in books such as *Electrical Power and National Progress* (1925), was a proponent of corporatist, planning solutions to social problems.[17] Other intellectual sources were J. B. Priestley's *English Journey* (1934) and a series of articles about the changing pattern of the English countryside in the *Architectural Review* by the architect W. A. Eden (1935).

Architectural modernism has a prominent place in *The Face of Britain* and a significant component in Rotha's network was the architectural community. Rotha was associated by early 1936 with the architect Basil Ward, who was partner in a practice very closely identified with the introduction of continental modernism into Britain (see Peto & Loveday 1999: 19, 128). When he established his trade organisation Associated Realist Film Producers in 1936, Rotha made Ward one of six advisors (see chapter three). Rotha also advised the public relations committee of the Royal Institute of British Architects (RIBA), of which Ward was a member, on a programme of films (see Haggith 1998: 46–52).[18]

The Face of Britain: dialectical documentary

In *The Face of Britain* Rotha took up both parts of Eisenstein's commitment to dialectic, as a mode of historical account, and as a mode of film construction at both microscopic and macroscopic levels.[19] 'Heritage of the Past', the first of four sections and analytically the film's thesis, paints an idyllic view of Britain's landscape and community before industrialisation. 'The Smoke Age' provides the antithesis, describing the impact of the industrial revolution on the land and people of Britain. 'The New Power', the synthesis, introduces hydro-electric generation. 'The New Age', in the rolling dialectic the new synthesis to the foregoing two sections redefined as thesis and antithesis, demonstrates the planning of Britain, which the film argues electrification permits.[20] The more microscopic use is particularly visible in a 'smoke age' section; rhythmically, at the end of the swing of a workman's shovel, Rotha first intercut several views of the industrial landscape of the Potteries, then a graveyard. The synthesis comes as the rhythmic soundtrack ceases, with a shot showing a graveyard's funerary statue in the foreground and smoking factories in the distance. In the midst of his first rough edit, he wrote in a letter, referring to the alternative title of Eisenstein's *October* (*Oktyabr*, 1928):

> I believe, but god what modesty, that this *Face of Britain* is taking up where *Ten Days that Shook the World* left off with ideological documentary. It is the

Figs 2.2–2.4 Rotha's use of dialectical montage in *The Face of Britain*: working and living conditions leading to premature death

> first pure dialectic documentary to be attempted in this country. It is no longer just simply building something – tracing a thing from its beginning to a logical conclusion – it [is] stating a definite dialectic argument and asking the intelligent answer from the audience.[21]

It is clear from the example of this film that it was eminently possible for filmmakers on the left to take and leave various aspects of Marxist theory; to be persuaded of the truth of a dialectical outline to history but without necessarily choosing explicitly to represent the class struggle that was for many a core component. Rotha's choice, as we shall see, was to resolve the dialectic in terms of science, industry, technology and planning. He argued that 'we can see with little effort [dialectic's] possible application to the film, both in approach to subject and in technical construction' (1936a: 235). In other words, he was more concerned for an account that made good *cinematic* sense than with conducting an argument in the fine detail of Marxist theory.

The Britain represented in *The Face of Britain* is seen as the product of historical forces – notably of industrialisation – and which was amenable to further modification by the planning of town and country. The theme is introduced in 'Heritage of the Past'. Ninety seconds of music and birdsong accompany shots of harvesting before the commentary starts: 'Most of us, at some time in our lives, have looked back to the time when the face of Britain was beautiful and the natural products of the soil were harvested so that man might eat and live.' Church bells and organ music, combined with slow cutting between shots of Cuddington, a Buckinghamshire village, introduce the next passage of commentary:

> Most of us have longed for a return to that epoch of serenity, when, in the quiet villages each cottage and house stood in its garden, in groups around the church. In those villages, men lived and died with little thought outside the life of their self-contained community, loving and knowing only the things that belonged to the soil, which was their livelihood.

In other words, the thesis of the film posits a pastoral golden age, recent in historical terms, against which the arrival of industry was dramatic and revolutionary.

The film represented the industrial revolution in 'The Smoke Age' as both aesthetic and social catastrophe. Rotha described his cinematic intentions for this section: 'in direct contrast with the first sequence, the character of this is violent, powerful and aggressive, most of it being swiftly cut and intercut with conflicting movements'.[22] The film presents industrialisation, following the discovery of Britain's six major coalfields, symbolically as a storm:

speeded-up cloud movements, trees bending in the wind and underexposed footage of landscape, intercut with smoking chimneys, symbolise the intemperate arrival of industry. In contrast with the sunlit harvest and village shots of the first section, the audience is told that 'the sun was hidden behind the smoke cloud of furnace and factory', introducing a series of light metaphors to the film. The narration argues that 'without plan or order, without thought for future decades, industry was developed as fast as men could work and build'. This representation of industrialisation as rampant *laissez-faire* was typical of historians of the Fabian school. Here it is in contrast both with the 'pattern' (for which read 'order') of the past, and with the town and country planning suggested in the film's synthesis. The disorder of industrialisation had direct aesthetic consequences: 'There grew up huge centres of manufacture and countless small buildings uglier still to house the servants of industry. Meadow and field were blotched with giant tips for slag. The power of steam and coal dominated the land.' This follows the Potteries sequence, ending on its graveyard and industrial scene. The social impact is spelled out in a contrast in the next block of commentary. Industrialisation, it states,

> gave Britain a new place in the sun. It gave her industrial, economic and political power, but at how terrible a price in the degradation and destruction of human life. And so today we endure this heritage of the industrial revolution, spreading its congestion of factories and slums over the face of the land, leaving for new generations the shell of a prosperous age.

The screen is occupied, after the storm sequence, with a series of industrial landscapes, cut together in rapid sequence: steelworks, the Potteries, the Durham coalmining town of Shotton.

'The New Power' starts in the same lyrical mode as the first, with birdsong and water sounds slowly replaced by electrical whines. Shots of Highland streams and lochs are intercut with high voltage electrical apparatus. 'It will be seen', Rotha wrote, 'that the sequence works up from comparative peace and calm to a climax of flashing light':[23]

> So electricity is born to bring back the sun to Britain. A nation plan is formed to carry power and light into the furthest corners of the land. Lochs are dammed to make giant reservoirs of energy. Waterfalls are harnessed, rivers diverted along new manmade channels. Tunnels are bored and aqueducts constructed so that turbines may turn to create the great new driving power and enable huge quantities of this energy to be controlled by one man.

Fig 2.5 *The Face of Britain*: 'The New Power' – hydro-electricity promoted as the route to modernity

Rotha here attaches his light metaphor to a 'nation plan': 'From the sources of this energy to North, South, East and West, the pylons carry their living load, over mountains, fields and rivers, never checking in their stride as they carry the new power to the waiting cities and eager countryside.' By associating himself with the Modernist champions of the national grid, Rotha also aligned himself with Modernist aesthetics (see Matless 1998: 52); for Rotha, fitness for purpose – a Modernist definition of design – is the yardstick. The commentary that closes this section of the film concludes:

> ...the heavy smoke clouds of the past can be dissipated forever. But a plan for power is only part of the greater plan to make Britain a land designed for living in, to make a New Age possible. A great new world lies ready to be created. There is much to be done.

The last two sections of *The Face of Britain* conform to the CEB's public relations strategy of stressing the modernity of electricity. 'The New Age' section argues that planning, married to electrical power, could remedy 'the ghastly squalor, brought about by the uncontrolled spread of industry'. It returns to both the aesthetic and the public health aspects: 'Not only must slums be cleared, but we must see that nothing ugly takes their place. The new communities must be planned in whole and in detail with a full understanding of the cultural and practical needs of a new society.' The film's association of electricity as a social transformative force may also have been read in some quarters as Leninist, recalling his equation that 'Communism is Soviet power plus the electrification of the whole country.' Describing his intentions for 'The New Age' section, Rotha stated: 'An orderly march of progress, firm and dominating, a building up towards prosperity and optimism characterises this final sequence. It must be impregnated with an urgency to get these things done or else England will return to the chaos of the Victorian era of dirt and smoke.'[24] Here his language became more explicitly modernistic:

> Strange new architectures arise to meet the demands of a changing civilisation, shapes and forms of simple beauty, dictated only by the purpose which they are meant to serve. Out of steel, glass and concrete, the architects and engineers must transform the face of Britain.

The aesthetics of fitness for purpose were combined here with a stress on new technologies in a combination typical of contemporaries advocating planning-based responses to national problems. Here new technologies, rationally applied, are presented as the solution to old technologies introduced 'without plan or order':

> All the efficiencies and amenities of the twentieth century are ready to be used. New sources of power, new means of communication, new methods and new processes are here for the service of man. This is an age of scientific planning, organisation of cooperation and collective working.

This passage marked an explicit statement of the importance of modernity in Rotha's view of the world: hydro-electricity, novel modes of transport and processes are brought together under the umbrella of planning and collectivism. The language of planning characteristic of the 1930s was widely used, adjectivally extended, to speak of economic planning, state planning and, as with Rotha's usage, 'scientific' planning.[25]

Rotha here advocated a model of expertise in which the professions of architecture, design, engineering and science, employing new technologies of electricity and communication, were joined in an ethos of rational planning. Harold Perkin's account in his book *The Rise of Professional Society* characterises expertise as a type of property traded by the professional classes. He argues that 'most professional expertise does not enjoy a natural scarcity, and its value has to be protected and raised, first by persuading the public of the vital importance of the service and then by controlling the market for it'

(1989: 378). This film is an example of such public persuasion. Perkin says that 'above all, [expertise has given the professional] the psychic security to press his own class ideal, his own view of what society should be and how it should be organised upon the rest of society' (ibid.). Barry Barnes has discussed relevant terminology; he illuminates how in these terms, what Rotha advocated here was not so much 'technocracy', which is rule by experts, as what he terms a 'decisionistic' society, in which expertise is drawn upon but is not in control. In this model, these experts – 'scientists, technologists … administrators, lawyers, economists and the like' – have their main function in advising 'policy makers and decision takers, often elected politicians'. Their 'technical knowledge and skills gives them a privileged position in society, and a real, though subservient involvement in decision making' (1985: 99-100). This suggests a three-tier society of political elite, experts and the general public. This last group, the argument goes, are cut-off from any real involvement in politics because access to power is mediated by possession of knowledge, which they are denied. Rotha's determination to use documentaries to build an informed citizenry was therefore a matter of political belief. At its conclusion, his film espouses a model of a democratic society making full use of expertise:

> Britain has tremendous wealth of human material and physical resources. She is now at a turning point in her civilisation. If her citizens will realise alike the opportunity and their own responsibilities, they can make this ancient land a well ordered and gracious heritage where the sun will always shine on the children of the future.

The use of the citizenship motif in the hands of the historical materialist Rotha bears a different cargo from that found in the more politically centrist documentarists. Just as the dynamic of history was, for him, forging new versions of nationhood, so the urgency of citizenship rhetoric was not for citizens simply to celebrate modernity, but to become actively engaged in further transformations. With these words, we see the completion of the cycle of light metaphors. At this point – in a symbolic visual image that he very soon regretted (see Rotha 1936a: 228) – the audience sees a couple walking on the crest of a hill. As the final words are spoken, the man points at the scenery with outstretched hand. A summary, perhaps prepared as a press release, stated that the film 'tries to use the documentary form for the expression of a dialectical purpose – to set out an argument plainly and simply yet allow the audience to draw their own conclusions'.[26] It is clear that the only conclusions available to the audience were to support the technological, expertise-driven, planned society it proposed.

Fig 2.6 *The Face of Britain*: 'The New Age' – an emphasis on modern architecture and scientific planning

Documentary modernism

The Face of Britain is a particularly striking combination of themes characteristic of 1930s modernity, conveyed using Modernist technique. It is replete with references to novelty, associated with technology – especially with 'the new power' of hydro-electricity – which usurps the natural with the man-made. Technology is also associated with planning. If this film were the sole example of a documentary singing the praises of the transformative power of science and technology, it would be an interesting curiosity. But many documentaries represented such themes and used varieties of Modernist technique to do so. It has become a commonplace that 'if modernity is shaped by technology, then the converse also holds: technology is a creation of modernity' (Brey 2003: 33). But science and technology as agents of change rarely occur alone in these films; rather they are elements within composites that more generally celebrate rationality and rationalisation, including the state and its bureaucracy. It is in this composite modernity that their interest lies, where their particular modernism can be found.

The language of 'modernity' and 'modernism' is notoriously slippery (see Cooper 2005: 113–49). 'Modern' and cognate terms, deriving from the Latin root *modo*, relate to ideas of currency and to novelty, to distinctions between the characteristics of the present and those of the past. From the nineteenth century onwards, the connotations of 'modern' have tended to be positive, unlike earlier usage (see Williams 1976: 208–9; Childs 2000: 12). The documentarists established the currency – modernity – of their films in part by representing novel technologies, rationalisation and 'scientific' approaches, and the transition to the better ways of living that they saw them as enabling.

Scholars in both the humanities and social theorists have been concerned with the transition to modern economies, mentalities and culture. Epistemological accounts of modernity typical of the humanities often place stress on the transition from an organic to a mechanistic worldview associated with the rise of science. Some authors locate this in the Renaissance, others in the Enlightenment of the eighteenth century. Social theorists including Marx, Weber and Durkheim, by contrast, have stressed processes and institutions such as industrialisation, capitalism and rationalisation as core to 'modernisation', defined as the emergence of modern societies (see Brey 2003: 35–40). This is the tradition that Rotha found himself within, probably unwittingly, in *The Face of Britain*. For people in this tradition, the industrial revolution was often the core event.[27] Both traditions describe modernity as a historical period stretching from whichever origin they favoured up until their present and into the future that the processes they described were producing.

Compared with the disputes over modernity, there is more general agreement over the term 'modernism', used in a broad sense to refer to responses of all kinds to modernity, and in a more specific sense – which I will term here 'aesthetic modernism' – to denote Modernist artistic movements. In this sense, present from the mid-nineteenth century, artists responded to their perception of the novel state of modernity by rejecting realistic conventions for artworks and substituting a concern with formal characteristics (see Brey 2003: 36). Raymond Williams discusses two traditions, with different histories. First, he speaks of 'the Romantics' victorious definition of the arts as out-riders, heralds and witnesses of social change' (1992: 24) under which the innovations in social realism of the mid-nineteenth century were core to the Modernist canon. But he explains how a later generation has usurped 'modernism' for itself: 'Futurists, Imagists, Surrealists, Cubists, Vorticists, Formalists and Constructivists all variously announced their arrival with a passionate and scornful vision of the new' (1992: 25). He identifies the period we are discussing, c.1890–c.1940, as the period of the 'absolute modern' and links it not only with the revolution in the media of cultural production brought about by photography, cinema, radio and latterly television, but also with the association that participants made with Freud, the subconscious and unconscious, and the associated questioning of the process of representation. Architectural modernism, of all aesthetic modernisms, most often embodied an aesthetic of gleeful response to the modern age; Le Corbusier is only the most obvious example, with his paeans in *Towards a New Architecture* (1927) to ocean-going liners and aeroplanes, and his prescription that a house be a machine for living in. Under the same terms that any of these artistic movements may be described as examples of aesthetic modernism, so may Soviet montage theory. As a response to the newness of post-Revolutionary society, with an emphasis on the cinematic form of the film work, and on the effect of that form, it had kinship with futurism, constructivism and other Modernist artistic movements, despite the repudiation of modernisms and the reign of socialist realism under Stalin (Murphie & Potts 2003). It is part of the diversity found in British documentary that, in addition to dialectical montage, they not only represented modernity but also adopted many of the modernistic aesthetic modes then current. Ian Christie has commented that 'in effect, documentary became an "applied" avant-garde, inheriting much of its theory and practice from the avant-gardes of the previous decade, whilst recasting these in instrumental rhetorics of information' (Christie 1998: 451).

Modernist thinking has close affinities with what David Bloor, following Karl Mannheim, named the 'Enlightenment' style of thought:

> Because Enlightenment thinking is often associated with reform, education and change it tends to have a strong prescriptive and moralising flavour. It is not meant to be the vehicle for neutral description, but a way in which a reforming 'ought' can be made to confront the recalcitrant 'is' of society ... The abstract universalism of the Enlightenment style enables it to hold up clear, general principles whose very distance from reality can serve as a reproach to the latter and a goal for action. (1991: 62–4)

At the level of abstract description we have here a way of talking about the documentary representation of science, technology and medicine. As citizens of modernity, documentarists represented its emblematic sciences and technologies as key components of modern life. Writers on early cinema agree that cinema was of a piece with modernity (see Gunning 2005). But some documentarists went beyond merely representing these themes, as their use of modernistic technique involved a reflexive self-consciousness and formalism that the older 'interest' films did not possess.

It is necessary to move beyond the individual case study, to establish how typical *The Face of Britain* was in its representation of science, technology, rationalisation and public health. Like those seeking a *rapprochement* between the small-scale case studies typical of the history of technology and the generalisations of modernity theory, we need to take the general conclusions of our analysis and see how they relate to other examples.[28] The earlier detailed analysis of the film revealed four aspects that might be related to broader cultural phenomena of the era. The film's ideas partake of a commitment to new technology, scientific planning and architectural modernism (that derive respectively from Quigley at the CEB and Rotha's associates on the Modernist wing of RIBA); a Marxist dialectical account of history specifying industrial technology as key to the transition to modernity (by way of Raymond Postgate); and a state of the nation discourse typical of the Depression era (originating from Priestley), which also accepted the dominant account that the health problems of the present were the products of industrialisation in the past. These ideas, as we have seen, were conveyed via a Modernist technique of dialectical montage (drawn from Eisenstein). It is possible to understand Rotha's film because of the unusually rich contextual material preserved in his archive, in the coincident composition of *Documentary Film*, and his own autobiographical account (1973). The following analysis moves from the general features of *The Face of Britain* to the particular examples of several other films. This analysis, explaining how a wider range of films represented these same and related themes, will make the case for the importance of science, technology and rationalisation as major aspects of documentaries made in this period.

The Coming of the Dial

Stuart Legg's film *The Coming of the Dial* (1933) contains a very explicit commitment to technology as applied science. This perhaps unsurprising for a film with the job of promoting the automation of telephony by means of the introduction of dial telephones and exchanges employing banks of electromagnetic relays in place of human operators. Right at the start of the film, there is a powerfully Modernist juxtaposition of thirty seconds of shots of one of the light display machines created by László Moholy-Nagy, with a commentary that intones:

> Research: the creative power behind the modern world. Building the future in the laboratory. The industrial chemist determining a carbon percentage for safety steel. The physicist analysing coloured light rays for signal lenses. The plant-breeder pollinating selected grasses for mountain pastures. These men are applying the laws of science to everyday problems. And research into the behaviour of electromagnets has revolutionised the telephone system and introduced the dial.

The juxtaposition of these words with the artfully shot and edited light display machine has a powerfully Modernist aesthetic. But, although the film

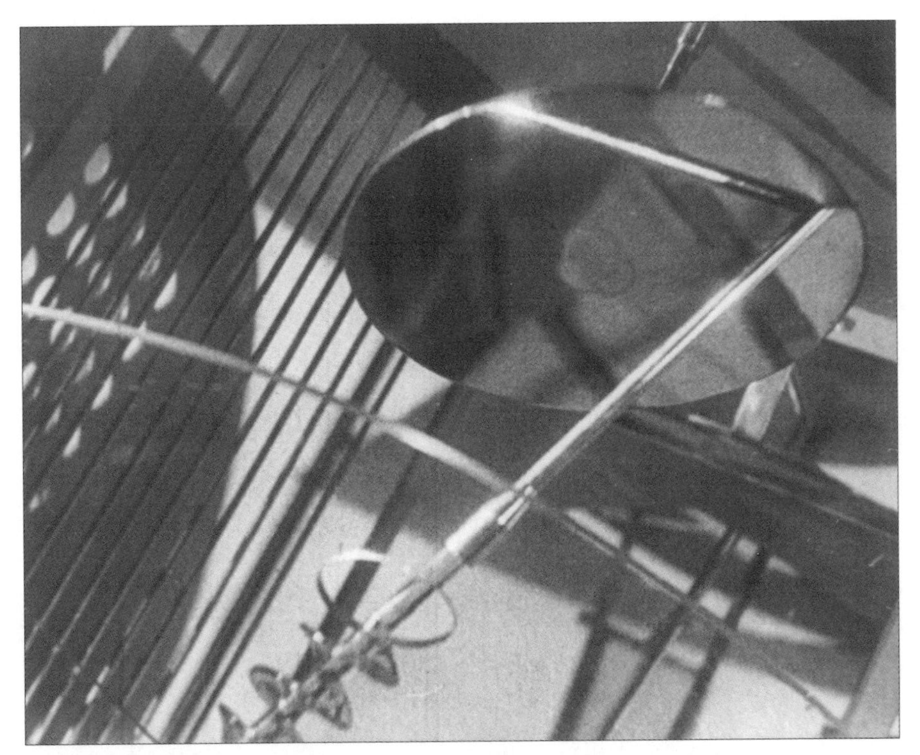

Fig 2.7 *The Coming of the Dial*: László Moholy-Nagy's light display machine

as a whole is conspicuously well made, with carefully angled and lit shots, cutting, use of superimposition of images and on the soundtrack electromechanical switches in place of music, it is not so visibly an example of dialectical montage as Rotha's slightly later film. But the presence of the work of Moholy-Nagy, a Constructivist artist who had the unimpeachable Modernist credentials of having led the Bauhaus, exemplifies another type of Modernist technique at the GPO Film Unit, namely the use of Modernist artworks in films. There was no necessity for Legg to incorporate a Constructivist sculpture in the film; this was included as a symbol, to attach Modernist connotations to a technological subject.

The words, taken by themselves, constitute a very bold assertion of the role of technology as applied science in constituting and accelerating modernity, notably in the phrase 'building the future in the laboratory'. But scientific research is here placed in the company of applied rationality more generally, notably in a sequence on the planning of telephone exchanges:

> [Ranks of telephone mouthpieces] The coming of the dial demands a new network of exchanges [superimposed sheets of numerical tables]

> to house the delicate mechanisms which convert impulses into telephone calls. These statistics, collected over years of telephone practice, tell the designer [close-ups of slide rule] how much conversational traffic each new exchange must handle and at what times of the day it will be heaviest. [Close-up of right eye of designer] So, he calculates the particular needs of each district [architectural drawing] in terms of traffic averages and telephone densities. On this basis, the plans for each new exchange are drawn. [Builder on scaffolding] So, the plans of the draughtsman take shape in steel and concrete. And new buildings appear against the London sky.

Science, technology, rationality, design and novelty are here all joined together. The film celebrates the process of modernisation in the context of 'modern business and modern commerce'; speaking of the construction of a new exchange in the City of London, the commentary states:

> Here, between St Paul's and the river, the old order is changing. Complexities of modern business and modern commerce demand the closest contact between office and warehouse, buyer and seller, wholesaler and retailer, stockbroker and investor, and between people who simply want to talk to each other. New methods replace old and this building rises above the city to act as the nerve centre of a new system of communication.

It returns to the centrality of science: '[the exchange] is the answer of the mathematician and the engineer to the modern demands for speed and organisation. Once more, the laws of science have been applied to an everyday problem and this dial system is a vital contribution to efficiency and speed.'

The Coming of the Dial shares with the final section of *The Face of Britain* a celebration of new technology as a component of modernity, but with more explicit links to science. It uses Modernist technique in its assumption of Moholy-Nagy's aesthetic. It is unlike the other film in that it contains no Depression-era reference to the state of the nation and no historical account. In these senses, it is much more of a piece with the Enlightenment style discourse that Bloor described; science is represented as a universal force acting and cultivated to improve the present.

Pett and Pott

The director Alberto Cavalcanti introduced another variety of modernism. Fresh from several years consorting with René Clair and other significant Surrealists in Paris, where he had made several films, including *Rien que les heures* (1926), Cavalcanti introduced this aesthetic into British documentary. His *Pett and Pott: A Fairy Story of the Suburbs* (1934), is in a whimsical surrealistic style. This used the technique, at this stage rare in documentaries, of employing an entirely fictional narrative. (This was one stage beyond the more usual documentary-drama technique of reconstructing an ideal-typical narrative based on real circumstances.) The grammar here was entirely different from the reportage and impressionistic approaches that Rotha had described. It constructed its narrative around the dialectic of the virtuous Pett family living in one semi-detached house and the sinful Pott family living in the other. The virtue of the Petts extended to the acquisition of a telephone which, in the film's climax, was used to apprehend a burglar in the Potts' household. No opportunity was missed to signify the contrast between the technologically aware Petts and the backward and degenerate Potts. At the film's conclusion, in a court scene played for surrealistic laughs, the modern virtue of becoming a telephone subscriber was driven home in a knowing and jokey manner. The judge summed up:

> It is possible, that at this stage of the proceedings, the court and public have drawn the inference that this case and the incidents appertaining thereto savour strongly of what I believe is commonly known as advertisement, on behalf of the Post Office telephone service. So it does! Nor need that fact in any way distress you, for we find, looking at the worldly affairs around us, that the good and just use and appreciate such telephone service while the evil and unjust ignore and despise it.

Scenes drawing on surrealistic conventions include commuters choreographed in synchronised swaying movements, and the superimposition of a train's whistle for a woman's scream, a reference to the Freudian unconscious typical of surrealism. Within the velvet glove of this film's surrealistic whimsy thrusts the iron fist of technological modernism. The film preaches that modernistic virtue rests in embracing the benefits of technology in the home, whilst turning away from it is aligned with vice and waste. The film proposes dialectically opposed models: in Marshall Berman's terms, the audience member has the choice to become either the subject or the object of modernity, to gain the power to change the world that is changing them, or to become its victim (see Berman 1988: 36). Compared with *The Coming of the Dial*, for example, modernism is transformed from being a remote 'continental' and academic pursuit to becoming an everyday choice for ordinary people; it has, I suggest, been domesticated.

Once again, the film is about the transformative power of technology in the present (which it shares with both the other films discussed so far) rather than its role in the past; this is a film which, like *The Coming of the Dial*, exists solely in a Modernist present. Congruently with this, the Depression does not impinge on this modern suburban scene. Though fluently edited, this is not montage in the Soviet sense.

We Live in Two Worlds

Cavalcanti, who was by then head of the GPO Film Unit, directed *We Live in Two Worlds* in 1937, with a script by J. B. Priestley, the *Daily Herald* journalist, novelist, playwright and broadcaster. The film originated when Grierson invited Priestley to use leftover footage from an instructional film on Swiss telephony, *Line to the Tschierva Hut* (1937), to construct a new narrative. The films were made in collaboration with Pro Telephon Zürich, the Swiss telephone provider (see Brome 1988: 209–10). Priestley set out what the title means right at the start:

> Nowadays we live in two worlds. I believe that's why we all feel so bewildered. It's not easy to live in two worlds at once, especially when they seem to contradict each other. The first world is the one we're always reading about in the newspapers: the world of separate states and nations; of frontiers and passports and customs houses and armies; the limiting and quarrelsome national world. Now for the second world that we don't hear so much about. It's the growing international world of universal trade, transport, communications. It goes on developing itself, circling the whole globe in spite of all political disagreements, all national differences.

He takes the example of Switzerland, hemmed-in by national boundaries: 'No doubt it is all very charming and mediaeval; some of them still work as they always did.' As in Rotha's film, rural scenes of traditional agriculture with lyrical music are contrasted with shots of high technology, in this case the firing of a rocket carrying a telephone wire up the mountain:

> This line connects a climbers' mountain hut with the town below, with Zurich and Geneva, with the whole world ... You see that telephone line has taken us from the first world to the second. From nationalism to internationalism. From this map – all angry frontiers – to this, showing all the lines of transport and communications: roads, railways, air services and telephone lines. Switzerland looks very different now, doesn't it, in this second, and more civilised, world?

Then the film makes a transition from focusing on the internationalism of communications to a different modernistic trope, of hydro-electric power, as in *The Face of Britain*, and the modern world it enables:

> This Switzerland is part of that second and larger world, the international world, harnessing its lakes and rivers with concrete and steel. Harnessing its mountain torrents to generate electric power. Power! Power for its new factories. Power for the tools of industry.

It continues with a paean to the internationalism of technological modernity:

> The world of international trade, transport and communication. International world, linking Switzerland with England by the language of the air, by wireless across oceans and continents. By telegraph and telephone across all frontiers ... across all frontiers... This is the new world, international world of concrete highways from city to city, of steel tracks from frontier to frontier. The world of the air where there can be no frontier ... no frontiers ... The world that sets great ships upon the seas ... majestic new symbols of its power and pride. It recognises no bounds but those of the great globe itself and asks to serve all men, whatever their race and tongue, just as air and sea and rail serve all men.

The internationalism represented in *We Live in Two Worlds* is of a very particular stamp; it is concerned with the capacity of technology to override nationalist politics and revolutionise the world. It sits alongside two contemporary modes of internationalist science, both that promoted by government action on non-scientific grounds – in which science is an adjunct to diplomacy – and that which is promoted between scientists independently of governments (see Elzinga & Landström 1996: 3).[29] Equally, *We Live in Two Worlds* makes explicit reference to the militarisation of Europe; idealistically it presents communications as the escape route from nationalisms to, as Priestley says, a 'second, and more civilised, world'. This film shares a clear commitment to new technology and the ghost of a dialectical structure with *The Face of Britain.* As we have seen, it replaces Rotha's concern with the state of the nation with an internationalist view. Its montage is not, however, slavishly dialectical.

Health for the Nation

Health for the Nation (1939), directed by John Monck for the GPO Film Unit on behalf of the Ministry of Health, shares with Rotha's film a dialectical account of history, a stress on industrial technology, a view of health problems as deriving from the industrial revolution and a commitment to rational, state-based and technical solutions to those problems (see Boon 2004a). In this impressionistic documentary, the sparse, poetic commentary is spoken by Ralph Richardson. The film is divided into three dialectical sections: thesis (the way things were before industrialisation), antithesis (the impact of industrialisation) and synthesis (how the health services are solving the resultant health problems). The thesis starts with shots of the timeless English countryside and a description of farming life. The antithesis shows the industrial development of the country, an impressionistic 'English Journey' accompanied by industrial location sounds, introducing the coal, iron and steel districts of England as well as the textile industry and transport. It states that 'out of iron and coal and steel we've built … slag heaps and smoke, soot upon the fields, forests of chimneys. In a hundred and fifty years, we have changed the face of Britain. We have changed it forever.' This section is amplified by a sequence on 'The people', and the impact of industrialisation on their health. The catastrophic interpretation of industrialisation is then given a forceful expression in an impressionistic sequence of panning shots of industrial areas, accompanied by the musical score at its most sombre. The voice-over states: 'Overcrowded, poor, under the shadow of disease. Into filthy hovels, into ill-ventilated factories and mines was crowded the manpower, the driving force of industry: men, women and children.' The synthesis is introduced with a sequence of the dates and titles of Public Health Acts, culminating in the foundation of the Ministry of Health. The film builds on this with a series of comparisons between the nineteenth century and the twentieth: water supply and drainage, house building, refuse disposal, medical services, infant welfare, school meals and milk, the school medical service, National Health Insurance, pensions. The concluding sections give an upbeat account of progress in responding to the health problems of the previous century:

> They're cleaning up our cities and our towns. We're stemming the tide of dirt and disease. We are safeguarding our men and women and children in their homes. We are modernising our factories, making them healthier places to work in. Air, light and labour-saving devices, clean water, refuse and sewage disposal. Clinics for mothers and children, milk and feeding. Hospitals for care of the sick, health and pensions insurance, new houses, homes for the old and infirm, these are the health services of the nation.

Health for the Nation, as much as *The Face of Britain*, shows technologies being used to overcome the problems of the past, a synthesis that enables the Ministry of Health to come across as truly modern.

Transfer of Power

Jack Beddington, who had commissioned Rotha to make *Contact*, established an in-house film unit at the Shell petroleum company in 1934 under Edgar Anstey, who worked at the EMB and GPO units. Arthur Elton, another veteran of the government units, became adviser in 1936. Shell's unit quickly established a reputation for making scientific and technical films. *Airport* (1934), produced by Anstey and directed by Roy Lockwood, with music by Jack Beaver, provides an interesting comparison with *The Imperial Airway* from a decade before. Although virtually all the subjects represented in the earlier film are also present here, the style is dramatically different so that, with rapid cutting, the subject matter is conveyed with much more pace, the style echoing the technological theme. Emblematic of Shell's output was Geoffrey Bell's film *Transfer of Power* (1939), an account of the development of the gear wheel. This example is illuminating because, although consistently referred to as a scientific film, its relationship to documentary is disputed. Rotha argued that 'this side of the Unit's work did not really fall within the category of documentary purpose, rather they were well-made films of instruction and explanation' (1973: 221). Such a comment illustrates the difference between a genealogical definition, under which this would be a documentary (because it was produced by Elton) and a definition based on its subject concerns or social interpretation. Counted as a documentary, it would belong to Rotha's reportage model, 'an honest effort to report, describe or delineate a series of events, or the nature of a process' (Rotha 1936a: 225).[30]

The film exemplifies what we may call a technical film aesthetic. This resides partially in its conspicuously high production values; it has carefully lit and live-action shots – including several showing industrial archaeological sites and historical artefacts from the Science Museum collection, historical illustrations and animated diagrams – enlivened with short phrases of tuned percussion music – crisply assembled to illustrate the commentary. The precision, authority and seriousness of the commentary are also essential to this aesthetic; unlike *Secrets of Nature*, there is no concession to humour in this film; it is designed to convey essential information only. This begins right at the start of the film:

> [Woman operating water pump] When people want a simple machine, they use a lever. [Rowing boat] People use a lever to push a boat along quickly; the oar is a lever. [Ancient stone relief] To lift heavy weights, people have always used levers as crowbars. [Shaduf] It was easy to raise water with a lever set up like this: the heavy stone balances the weight of a bucket and lightens the work. [Animated diagram] This simple bar working on a piv-

Fig 2.8 ***Transfer of Power*****: clarity of exposition**

> ot was the start of great things. When the pivot is in the centre, a weight at one end is balanced by an equal weight at the other end…

The film contains an idealised history of technology, but the style of this account is technological evolution, not historical dialectic. For example, showing first images of raising water using a simple crank windlass, then a four-handled windlass from Agricola's *De Re Metallica* and then an animated diagram, it speculates 'increasing the number of arms on the windlass may have led to the toothed wheel'. The story of the film takes us from here, by way of windmills and watermills, via the industrial revolution to synchromesh gears and (unsurprisingly for Shell) modern lubricants. It takes us from crude wooden gears in windmills to metal gear wheels, via technical explanations of both epicycloid and involute gear profiles.

The historical account of the adoption of steam power follows a conventional necessity-and-invention narrative:

> The windmill was the typical machine of the Middle Ages; wooden gears were good enough for its slow speed. Through the labours of craftsmen, it was efficient in its own time. But a new age was coming. People began

to work in metal. Mines had to be dug deep into the earth; water had to be drained from the mines. Crude and cumbrous water engines were used, but they were not strong enough to drain the deeper mines. Steam was harnessed to serve the new need … The age of wind and waterpower and the simple wooden gear wheel was coming to an end. The factory age had begun.

Unlike *The Face of Britain* or *Health for the Nation*, aspects other than the technical are not included in the account. In a small number of industrial sequences, the film does share some of the familiar visual and aural tropes; there is a concentration on men at work in industry and the soundtrack uses the 'wild track' of industrial noises and indistinct working speech typical of many documentaries, where filmmakers had not been able to afford the expense and difficulty of taking sound recording equipment with them.

New Worlds for Old

Several films Rotha made in the later 1930s could be used to illuminate how his style in representing technical themes developed from the genre established with *The Face of Britain*. *New Worlds for Old*, which he made for the Realist Film Unit in 1938 on gas industry sponsorship,[31] suits our purpose, despite his disavowal that 'it was a cod film and I was in no serious mood when I wrote and made it' (1973: 225). This manifests itself in virtuoso spoofing of several existing British documentary genres (see Low 1979a: 137–9), a rather arch script and some joke sequences. A score by William Alwyn acts in counterpoint to the film's humorous approach. Despite its tone, both its subject (which derived from Political and Economic Planning (PEP) reports on the electricity and gas industries) and its technique render it highly relevant to this narrative. On a Rockefeller Fellowship to New York in 1937–38, Rotha had witnessed the New Deal Federal Theater 'living newspaper' theatre shows, in which news stories were acted-out, often with a voice from the audience interrogating those on stage (see Pearson 1982: 73). He applied the idea – later known as multi-voice commentary – in this film by having several voices interrupt Alistair Cooke's main narration. At first, there is an unseen, sceptical voice and, towards the end, several interlocutors, who stand for particular points of view, are seen on screen. The interplay of voices serves to drive forward the film's argument, an alternative to the caption boards that denote the transition between dialectical sections in *The Face of Britain*.

The film commences with a historical prelude on the Victorian Age, using Tussaud waxworks, starting with Queen Victoria. An uncredited commentator, establishing the mocking tone, says: 'There she was. All of her.

The Queen of England.' The commentary continues, against shots of smoke stacks, 'a wilderness of chimneys was the Promised Land', followed by a quick résumé of nineteenth-century engineering landmarks. A burlesque sequence of nineteenth-century life follows. The Victorian age is represented as the gas era.

This, by now conventional, documentary stress on industrial technology is continued in the next section, which Rotha may have intended as an antithesis. Cooke assumes commentary duties here, stating that 'in the post-war years, all the marvels of science and mechanics changed our lives [man listening to wireless through headphones, aeroplane, cars on bypass road]. Prominent in this new age was the rapid growth of electricity with its gleaming apparatus, its symbols of a new era of power.' The remainder of the film may be seen as an extended synthesis comparing gas and electricity, both 'modern fuels' derived from coal. A sequence demonstrates how inefficient coal is in industrial and domestic use, with two-thirds of its value wasted. A Kentish miner, voiced in exaggerated middle-class tones, has to wait two hours for a coal-heated bath and a working-class couple have to tolerate the heat of a range in high summer. An interlocutor butts in: 'That seems very stupid … I thought we were supposed to be living in an age of science.' We see coal being delivered. The interlocutor says: 'You can't get coal out of a tap.' Cooke responds with: 'Oh yes you can', leading via a sequence on the chemistry of coal to the manufacture of gas. 'You just take a lump [of coal] into a laboratory and see how a chemist would treat it scientifically.' This involves a sequence in which three white-coated, bowler-hatted 'scientists', in operatic voices sing:

> It's dirty
> It's heavy
> And when you burn it to get heat it makes soot and ash and gives off smoke that dirties the air.
> But give it to us and we will turn it into clean, flexible, automatic heat for you.

Firstly, an animated diagram then a rather breathless gasworks tour explains the production of town gas, coke and by-products. The voice interjects: 'That's all very plausible and convincing in the laboratory and in the gas works and on the screen, but what about in real life as it affects all of us?' Use of gas for cooking at a Lyons Corner House, the Prunier Restaurant and the D. H. Evans department store is shown. Gas heating and air conditioning is exemplified by their use in a cinema. The first interlocutor seen on screen asks how use of gas can improve things in industrial areas with smoke

Fig 2.9 ***New Worlds for Old*****: 'just take a lump of coal into a laboratory and see how a chemist would treat it scientifically'**

problems. In response, Cooke explains its use in industry – in steel works, as a precision energy source in manufacture and in heating and typesetting at the *Daily Telegraph.* It is 'the new, scientific form of clean, automatic heat'. Another interlocutor asks about life for the working man and his wife in slum areas. Cooke picks up: 'You'd think that the housewife is the last person to benefit from scientific research, but gas is showing that a properly-run home needs the best form of heat, heat which is not only on tap but which can be controlled and regulated.' We see a 'working-class' woman in a gas-equipped kitchen and Mrs Bumble, a cook general, enthusing about 'all these contrivances that have come along … why it's just wonderful the way everything looks after itself'. A woman interlocutor complains: 'That's all very nice for those that can afford such luxuries, but it's no good to us poor people. Think what it will cost.' This is an invitation for Cooke to assert that the vast majority of subsidised housing postwar has been equipped with gas. A child hymns life in the new flats. Over shots of the gas industry's Kensal House flats, the voice directly raises the possibility that electricity – 'the more modern form of fuel' – might do all this better than gas. Cooke answers this with two international comparisons. First is the example of New Jersey, where many consumers choose gas for heat and electricity for light. Then, the example of

Switzerland is conveyed in a joke sequence in which the same shots are run twice, first voiced in French, then in English; despite its abundance of hydro-electric power, the Swiss are importing coal to make gas. The film ends with a passage of ironic hyperbole: 'those hard-won hours at the end of the working day can be spent as you like because this clean, scientific, controlled supply of heat has become part of your life and mine'; we see people knitting, doing jigsaws and tending gardens.

New Worlds for Old, for all the unusualness of its tone, represented science and technology, and praises the modernity of both gas and electricity. It also presented the public with a rationalistic analysis of a contemporary issue in the PEP mould, using multi-voice commentary, a technique new to documentary.

Modernity, technology and documentaries

The modernistic visual and aural components of all the documentaries discussed here convey a series of celebratory statements about modernity and technology, to which they are the aesthetic counterpoint. All are cinematically fluent, mostly in the impressionistic mode, though the three that employ a dialectical account of history are the same that use dialectical montage. Most are concerned with the state of the nation (and *We Live in Two Worlds* with the state of the international system). The two furthest separated amongst our examples are *Pett and Pott*, with its surrealistic version of documentary drama, and *Transfer of Power*, with its commitment to the lucid communication of technical information. The interest of the films discussed here to the historian of modernity would be as second- or third-order vehicles of modernistic ideas. To the film historian, they reveal the significance of modernity to documentarists. To historians of science, technology and medicine, they demonstrate the cultural context within which these subjects were understood at this time. Even if Cavalcanti may be mentioned in the same breath as that better-known Surrealist filmmaker Luis Buñuel, Rotha was certainly no Marx, Legg no Mumford and Priestley no Weber. All the same, these films demonstrate the pervasiveness and credibility of representing and promoting modernity and technology in the interwar period.

3 DOCUMENTARIES AND THE SOCIAL RELATIONS OF SCIENCE

> The willing co-operation of men of science, the financial support of persons belonging to different political parties (or to none at all) and the creative work of film directors with the outlook of Grierson and Rotha could be enlisted to quicken the social imagination of England. (Hogben 1936a: 8)

The contexts within which British documentaries were first made produced a form that had emphasised technology, applied science and technological modernity generally. From around 1936 there arose a new context that produced a fusion of documentary filmmaking, science and social concerns. The new focus for documentaries was possible not only because of widespread reactions to the social costs of the Depression but also because of changes both in the public culture of science and in the funding and ambitions of documentarists. Also significant was the associative culture of the 1930s, in which heterogeneous groups of individuals formed associations, committees and clubs, often with overlapping memberships, to pursue many kinds of quasi-political activity.[1] The phrase 'middle opinion' has been used to refer to the breadth of political opinion represented around common interests in such groups, including organisations such as Political and Economic Planning (PEP) and the Next Five Years Group (NFYG) (see Marwick 1964).

Changes in science: the social relations of science

During the 1930s, several groups of scientists actively debated the social relations of science. This was a reaction to perceptions both of science's culpability for rendering World War One particularly terrible with new weapons such as poison gas, and of the Depression as the product of technological unemployment, 'the march of the machine'. These factors led some scientists early in the decade to begin to doubt what had become an established positivistic association between the accumulation of scientific knowledge and the progress of mankind (see Collins 1981: 228–9). In addition, the growth of the sciences was leading to broadening ambitions amongst scientists for greater influence on society and its direction. This in turn increased activity in what Frank Turner has called 'public science', which had been a feature of science's place in culture since at least the 1880s:

> Scientists in their capacity as observers and interpreters of physical nature still remain part of the larger social order, and between them and it there exists a dialectical relationship of mutual influence and interaction. As one result of this interaction, scientists find that they must justify their activities to the political powers and other social institutions upon whose good will, patronage and cooperation they depend. The body of rhetoric, argument and polemic produced in this process may be termed public science. (1980: 589–90)

As some scientists broadened their ambitions, and as they worked with other professional groups including architects and filmmakers, they developed several intertwined élite discourses that made ordinary people their object and proposed how science might be made to impact on their lives. The application of science to the diagnosis and treatment of social problems in the 1930s certainly sought to liberate the poor but at the same time it did nothing to diminish the elevated social status of the scientists and other experts over the objects of their policy prescriptions. And, for all that, discussions amongst professionals were *about* the public, they were constructed in private, in closed meetings, film studios and editing rooms.

Historians have considered the interwar discussions on science and society on three levels: some have focused on the left political commitments of scientists, others on informal associations of scientists, and others again on more formal organisations, notably the British Association for the Advancement of Science (BA).[2] Julian Huxley features in all these accounts. We should note however that the 1930s 'turn to the social' covers several different phenomena. There was both concern to defend science against criticism for what some people saw as its negative effects on society and, simultaneously, advocacy of a larger role for scientists and scientific modes of thought in the conduct of society and government. Authors have reified all this diverse activity as a 'social relations of science movement'.[3] This account avoids this reduction and, instead, pursues the distinctiveness of particular examples.

One informal forum, typical of the period's associative culture, where the social relations of science were first discussed between 1931 and 1933 was the 'Tots and Quots', a group convened by the zoologist Solly Zuckerman.[4] This group gathered over dinner in London more or less monthly, with discussions led by invited speakers about the economic and social relations of science. Its membership was diverse; attendees included the scientists J. D. Bernal, Patrick Blackett, J. B. S. Haldane, Philip D'Arcy Hart, Lancelot Hogben, Julian Huxley, Hyman Levy, Joseph Needham, Gip Wells, the architect Godfrey Samuel and the economists M. M. Postan, Roy Harrod and the MP Hugh Gaitskill. Zuckerman recalled that 'our talks roamed over wider and wider issues, but more and more what we debated was the question of the general significance of science to society, and the conscious role science might play in social development' (1978: 394).

The meetings of the BA were another site of continuing discussions on science and society. There was a long-running debate about whether the BA's purpose should be to provide a meeting place for professional scientists or whether it should act as the public relations agent of science, so as to counter any public impression that the negative impact of the sciences outweighed their beneficial effects. Paul Rotha's associate Ritchie Calder, who covered science for the *Daily Herald*, had access to the BA's members and to some of its internal deliberations. In 1932, he criticised the BA because 'not a single great social issue had been faced by the British Association' (quoted in McGucken 1984: 36). But, after a series of interventions by senior scientists including the biochemist Frederick Gowland Hopkins and Richard Gregory, editor of *Nature*, a resolution was placed before the Association's general committee in 1933 requesting 'the Council to consider by what means [it might] assist towards a better adjustment between advances of science and social progress' (quoted in McGucken 1984: 40). In the first instance, at their 1934 meeting, this produced a new concern with applied science and its impact on the economy, as well as some discussion of science and economic planning in relation to agriculture and rural life (see McGucken 1984: 45). This change in policy can be seen as the resolution of the debate in favour of a public role as 'apologist of science' (Collins 1981: 228).

From 1936, under the impetus of concern over Nazi Germany (especially Aryan race theory) and Fascist Italy (notably use of mustard gas in Ethiopia), the BA began also to discuss international factors. These discussions led to the formation in August 1938 of a Social and International Relations Division under the chairmanship of Richard Gregory (see McGucken 1984: 100–6). Its purpose was, in the words of a *Nature* editorial, 'to further the objective study of the social relations of science. The problems with which it would deal [were] the effects of the advances of science on the well-being of the community, and, reciprocally, the effects of social conditions upon the advances in science' (Werskey 1988: 245). The political beliefs of the members may not have had such a bearing here as some members' explicit commitments might be thought to imply. The four more politically left members of the Divisional Committee – J. D. Bernal, P. M. S. Blackett, Lancelot

Hogben and Hyman Levy – were more than balanced by the wide range of political opinion of the rest of the committee, which numbered more than 35 (see Werskey 1988: 246). They were also outnumbered on the executive committee, which, with Gregory in the chair, included Ritchie Calder, the Director of the London School of Economics and member of the Eugenics Society A. M. Carr Saunders, Huxley, the mathematician Hyman Levy and the nutrition scientist John Orr (see McGucken 1984: 131). More than this, the grounds for agreement tended to exceed those for dispute. So, for example, we find Bernal writing: 'An attempt will be made to find out … what are the actual results of applied science in the contemporary world and what they might be if science were applied in a rational and ordered way for human welfare' (quoted in McGucken 1979: 255). This is very similar to Julian Huxley's frequent calls for the application of rationality in human affairs, for example: 'the expenditure of research funds primarily on industry, then on war, and finally – as if it were an afterthought – on psychology and sociology, is a complete reversal of the direction of human needs'; he further commented that 'we are almost forced into a Marxian view of the relation of science to the State' (1936b: 127, 126). This from the author who, within months, was trumpeting the non-party basis of the Next Five Years Group (Huxley 1936c: 17). But again citizenship discourse defused any political tensions – at least outwardly – as, in 1938, the BA 'concluded that [scientists] were responsible neither for the positive nor especially for the negative social uses of science. In a democratic society such uses, they plausibly concluded, were the responsibility of every citizen, scientists and layman alike' (McGucken 1979: 264).

We may draw firm conclusions about the impact of the 1930s on the relationship of scientists to society from all three kinds of account. As we have seen, there were new emphases both on the application of science and rationality – often in the form of planning – to society, and on the importance of science's relations with the public. These opened up new opportunities which documentarists, products of the same circumstances, were well placed to exploit.

Changes in documentary: sponsors, contributors and form

The funding of documentary filmmaking has consistently been precarious. Writing two decades after the events described here, Rotha summarised the situation:

> The machinery, and the commercial interests controlling that machinery, of theatrical-film distribution … have always been inhospitable to the ac-

ceptance of documentaries in the cinemas … As a result, the economics of documentary have relied almost entirely on a need-to-be-served, a purpose-to-be-met – in fact, sponsorship. That is no secret. (1955: 367)

After Stephen Tallents, Grierson's mentor and champion, ceased to be public relations officer at the GPO in 1935, it became more difficult for the Unit to make films on subjects other than GPO services. Treasury control of the clients and subject matter of GPO films had become increasingly stringent (see Swann 1989: 57–64, 76–8). At the same time, not only government departments but also industrial concerns (such as Imperial Airways and Shell and voluntary organisations (such as the National Council for Social Service) were increasingly persuaded of the public relations virtues of films. The impact on the representation of science, technology and medicine was significant. The Shell Film Unit, founded in 1934 and well known for its technical films, was typical in the way that it espoused indirect public relations; a memorandum from its head, Alex Wolcough, on the aims of their film programme listed its aims: 'To improve the efficiency of the Shell organisation by creating a greater knowledge of its products, and by teaching modern methods of marketing; to help improve the demand for Shell products; to create general goodwill with perhaps no immediate or directly traceable return' (cited in Gordon 1999: 2–3).

A significant new sponsor of documentaries was the gas industry, which in that period was comprised of many local companies, with national coordination in marketing and public relations provided by an organisation called the British Commercial Gas Association (BCGA). The film sponsorship activities of this industry give a particularly strong exemplification of the interpenetration of public relations and citizenship discourse, which had notable implications for the representation of science and rationalising expertise more generally. Each year from 1935, they sponsored at least one high profile film on a social subject designed 'to appeal to the public conscience' (De Mouilpied 1937b: 611). *Housing Problems* (1935) and *Kensal House* (1937) presented life in the slums and in the flats built to replace them, whilst *The Smoke Menace* (1937) covered smoke pollution from the domestic hearth and industry. *Enough to Eat?* (1936) – the main subject of this chapter, sponsored by the Gas Light and Coke Company (GL&CC) – discussed malnutrition in Britain. *Children at School* (1937) exposed the poor state of the nation's schools. *New Worlds for Old* (1938), as we have seen, treated the gas and electrical supply industries; and *The Londoners* (1939) celebrated the work of the London County Council in the alleviation of social ills, with particular stress on public health.[5] The gas industry's film catalogue, *Modern Films on Matters of Moment* (1939), opened with these assertions:

> The films in this catalogue are offered as dramatic accounts of some of the problems of modern Citizenship in which the general public and the Gas Industry have a common concern ... Nutrition, Housing Reform and Public Health ... Some outline a planned approach to home economics and some discuss the National problems of Smoke Abatement, Nutrition, Housing and Education. We offer this account of matters of moment in the life of the community in the hope that it will serve not only to make known the activities of the Gas Industry and the responsibilities which the Industry has taken upon itself in matters of Public Health and general welfare, but also help to articulate the public knowledge in the major social problems. (British Commercial Gas Association 1939: 3)

The fact that 'the problems of modern Citizenship' were presented as the grounds for discussion by the Gas Industry and the public confirms its importance as a common ideology in the creative alliances that produced documentary films. The point was expanded in internal publications: 'these films are part of an education in citizenship which will give to the ordinary man a better understanding of the social problems of the day. To do that is to give him a new conception of the Gas Industry. These two things are complementary' (De Mouilpied 1937b: 611). As the *Gas Journal* put it: 'This sort of film is good citizenship and good business too' (Anon. 1936b: 325). For the gas industry, these films were part of a sophisticated public relations strategy. Gas was in competition with the electrical supply industry, especially for large installations in the slum rehousing schemes of the local authorities. Clement Leslie, then Publicity Manager at the GL&CC, memorably recalled that the public relations problem of the gas industry was that local authorities 'associated electricity with the Millennium' (quoted in Rotha 1973: 155). And, as we have seen, electricity's main selling point was its modernity. Leslie took a different approach, asserting the modernity of gas by associating it with urgent public issues (see Swann 1989: 101), hence the film catalogue title '*Modern* Films on Matters of Moment'. This was a second type of modernism additional to the celebration of technological modernity described in the previous chapters.[6] David Milne Watson, Governor of the GL&CC, in a letter to the *Times* stated:

> We have found here a means of establishing relations of improved understanding and increased confidence between our industry and its public, and not least in the minds of those more highly critical and influential sections which are particularly difficult to reach through traditional channels of persuasion. (1938: 12)

This was partially achieved by manufacturing a reputation for laudable disinterest. For example, one of their own publications, speaking of *Enough to Eat?*, asserted that 'such a film, naturally, contains no gas propaganda whatever', a sentiment that was picked up in reviews in both the general and the medical press: the *Lancet* stated that 'the film is noteworthy for the complete and conspicuous absence of any reference to gas, light, or coke' (Anon. 1936l: 927). The wording is almost identical in the *New Statesman* and *Nation*. This is particularly striking in this case, as the gas industry clearly had an interest in the cooking of food.[7] The BCGA continued to use nutrition as a subject for their public relations, running a campaign in 1937 and holding an exhibition in 1938 (see Anon. 1938a). Covering malnutrition was only a step away from discussing cookery but, in the 1930s, it seems the gas industry could persuade many audiences of the selflessness of its concern. A related rhetorical move was the deliberate restraint of the treatment. Where the Eugenics Society film *Heredity in Man* (1938), as we shall see, sought to use the dramatic effect of showing the 'mentally deficient', the producers of *Enough to Eat?* stressed the film's moderation. The commentary states that 'its aim is to describe the existing situation, but not in terms of harrowing pictures of the worst sufferers from serious deficiency diseases; these it has deliberately avoided'. Reviews also picked up on this trope, which we may conclude was featured in the press release. The *Lancet* reported that 'the absence of overstatement throughout drives home the argument for a constructive national food policy' (Anon. 1936l: 927).[8]

Production of these films served both the commercial interests of the gas industry and the financial needs of both documentary filmmakers and scientists.[9] But the financial imperative is not sufficient to explain the particular subjects and the representational choices adopted for the films. We have seen that the social circumstances of the 1930s had an impact on the documentarists' worldviews and politics. There is no reason to assume that figures within the public relations departments of the sponsoring companies would not also have mingled social concern, and perhaps also political commitment, with corporate self-interest.

Increasing numbers of documentarists in the mid-1930s, and especially after the contraction at the GPO, joined Rotha in making factual films outside government, often at companies such as Strand (from 1935), Realist (1937 onwards) and Gaumont-British Instructional. From the point of view of these documentarists, when freed from the constraints of representing EMB or GPO services, they were at liberty to represent wider topics wherever they could find funding and like-minded allies. It was the enrolment of wider groups of individuals in the process, including some scientists, that nuanced the choice of subjects and their representation. The precise circumstances were significant. Given the business interests of the gas companies, we might expect films on domestic heating or lighting, for example, to dominate. But the relatively small number of such films is relegated behind the films concerned with social issues that dominate their catalogue *Modern Films on Matters of Moment* (BCGA 1939: 78).

The social problem documentaries produced with gas sponsorship had a particular focus on public health issues and their reform. It is significant that these films represented social problems as having solutions, often by the agency of corporations or government departments (see Swann 1989: 115–16). They promoted rational, planned and paternalistic responses to social problems. Grierson, characteristically stoking the fires of rhetoric, enthused in November 1938:

> In many of the documentary films, the country is shown tearing down slums and building anew, or facing up to unemployment and reorganising economically: in general passing from the negative to the positive. It is in this, precisely, that most of us have felt that the strength of democratic Britain is made manifest. (1966: 85)

Documentarists experimented with new styles of film to represent the new subjects. Rotha commented that 'the making of films about the task of social reconstruction was a great deal harder than the dramatisation of the steelworker and the fisherman as symbols of labour' (Rotha 1952: 196, written for the 1939 second edition). Unlike the majority of films in the previous chapter, which adopted an impressionistic style, the new films experimented with variations on the reportage mode: *Housing Problems* confronted the audience with slum dwellers speaking to camera; *Enough to Eat?* adopted an illustrated lecture format; and *The Smoke Menace* borrowed its style of reportage from the American newsreel *The March of Time* which, unlike other newsreels, was known for analytical treatments of political and social issues (see Fielding 1978). In all these examples subject and style were very tightly bound together, as is clear from Grierson's later recollection:

> We worked together and produced … the poetic documentary. But … there has been no great development of that in recent times. I think it's partly because we ourselves got caught up in social propaganda … got caught up with the problems of housing and health, the question of pollution … We got onto the social problems of the world, and we ourselves deviated from the poetic line. (Quoted in Sussex 1975: 79)

However, as Brian Winston persuasively argues, representing the victims of the Depression meant substituting 'empathy and sympathy for analysis and anger' (1995: 47). The concomitant prescription of proposed rationalisation and action on the part of the state merely distanced the audience from any need for action (see Winston 1995: 40–7).

The diversification of sponsorship and production facilities was accompanied by other consolidating activities: with *World Film News*, set up in 1936, documentary had a publication from which it could marshal its allies and tease its political opponents (see Swann 1989: 111, 77).

Scientists *and* documentarists

Filmmakers were assisted by Rotha's formation of an organisation named Associated Realist Film Producers (ARFP) late in 1935 to broker contacts between documentarists and potential clients. Its advertising leaflet stated:

> [The ARFP] has been founded as an authoritative body qualified to act as a consultant film organisation to Government Departments and other official bodies, to the various Public Services, University and Education authorities, Industrial and Commercial organisations, and others anxious to make their activities known to a wide public. Such bodies are often at a loss to know how to set about the making of a film which will be worthy of their purpose, which will not involve them in unforeseen costs and which will reach the audience they require.[10]

The foundation of ARFP may be seen as the third element (along with the publication of *Documentary Film* and the release of *The Face of Britain*) that made these few months particularly significant for the role of Rotha within our story. The membership of ARFP contained all the most prominent documentary filmmakers but, significantly, the organisation also had a panel of advisors, including the liberal and left biologists Julian Huxley, J. B. S. Haldane and Lancelot Hogben, along with the architect Basil Ward, the designer McKnight Kauffer and the composer Walter Leigh. The cinema trade press announced: 'Plan for non-fiction films: Professors to act as advisors' (Anon. 1936f: 22).[11] We may see their collaboration with documentary as an extension of the ideological base of the documentary group. An extended network of interests, ARFP was typical of the associative culture of the 1930s. We may note for example that both Haldane and Huxley had been on the editorial board of the *Realist*, 'a journal of scientific humanism', which ran between 1929 and 1930, publishing articles across the scientific, political and artistic spectrum.[12]

Looking at the activities of these individuals reveals the mutual benefits of this organisation for documentarists and scientists.[13] Each of the three scientific advisors to ARFP took a more or less active role in relation to films; Haldane and Huxley had both been founder members in 1925 of the Film Society and Huxley at least attended frequently (see Samson 1986: 310).[14] Hogben presented documentary in his publications and lectures as the means of achieving an education for citizenship that stressed the instrumental potential of science, as explored below. Huxley, as we shall see, had a deep commitment to film. He spoke about the potential of films in a lecture series at the Royal Institution in January 1937.[15] Haldane appeared in *The Smoke Menace*. Huxley and Haldane were amongst the liberal and left figures on the unlikely General Council of Kino, the film distributors associated with the Communist Party (see Ryan: 1980: 61).

Hogben's View

Hogben's interest in both scientific citizenship and visual communication make him a key figure for this account, articulating a citizenship discourse that linked documentaries and the social role of science. He argued that 'the cultural claims of science rest on the social fact that the use and misuse of science intimately affects the everyday life of every citizen in a modern community' (1937: 119). That translated into a responsibility for the scientist; 'this nonsense that the scientific worker has no time to be a socially responsible adult, exercising his social responsibilities as a citizen, is due to be debunked' (Hogben 1938: 11). His review of books by Rotha and Grierson in the film journal *Sight and Sound*, published at very much the time that ARFP was established, can be seen as his manifesto for the organisation. For him, the cinema – 'the university of the future' – had a particular purpose in relation to science, to sweep away the mystification normally caused by popular science, and replace it with a functionalist view:

> To make mankind aware that science teaches the possibility of plenty of the people by the people and for the people, we have got to translate it into the vernacular ... Education for the age of plenty means showing people that science is not a mystery. It is organised workmanship. This can only be done if you show people how science has arisen out of the common experience of mankind to meet the common needs of mankind. (1936a: 8)

He used Grierson's phrase, bringing 'the new world of citizenship into the imagination' to describe what he saw as the work of documentary (Rotha 1936a: 5). This can be seen in light of his views on science education in general:

> Education for citizenship demands a knowledge of how science is misused, how we fail to make the fullest use of science for our social well-being, and, in short, a vision of what human life could be if we planned our resources intelligently. (1937: 123)

This explicit linking of 'science for our social well-being' and planning became a keynote concern of scientists discussing the social relations of science, for example in the British Association's new Division. This scientific component sat very comfortably not only with Rotha's stress in *The Face of Britain* on technology, architecture and planning but, as we shall see, with Huxley's in *Enough to Eat?* Hogben, unlike Huxley and Haldane, never appeared in a film, but his clarification of the roles of science and documentary exemplifies how scientists' film appearances could be seen. He wrote to Huxley about *Enough to Eat?*, 'I want to tell you that I think your part in the Nutrition film was a really courageous piece of active citizenship', emphasising the stress on this political discourse in his version of the social relations of science.[16] But we may generalise about such appearances. At the minimum, we can argue that for the individual scientist to appear in a film implies assent to preliminary ideas about how its subject is represented. Beyond this, they assent to add to it a negotiated contribution of their own: a statement of opinion, a display of expertise, or the presentation of information.

Julian Huxley: a public scientist at the cinema

Julian Huxley was one of the most prolific public scientists of the mid-twentieth century. A regular in the magazines and newspapers of the day, he became a familiar radio broadcaster (see Huxley 1934a). With *Enough to Eat?* he also became the first public face of scientific filmmaking. Huxley's career in popularisation began in earnest when he resigned after two years from his Chair in Zoology at King's College London in 1927 to co-author the work *The Science of Life* at H. G. Wells' invitation. As a scientist, he is known for his work on ethology, especially bird behaviour, and laboratory work in ontogeny (the development of organisms). His work in evolutionary studies reached its peak with *Evolution, The Modern Synthesis* in 1942. But, apart from a few full-time posts, including Secretaryship of London Zoo between 1935 and 1942 and being the first Director General of UNESCO from 1946 to 1948 (see Baker 1976: 233–4), any scientific research he pursued was conducted in parallel with a high level of activity as an essayist, populariser, organiser and spokesman of science.

Huxley fits Frank Turner's definition of 'public scientist' in the sense that his compositions and performances all assumed that 'science is worthy of receiving public attention, encouragement and financing' (Turner 1980: 590). But, as a rationalist he went beyond the older concern with supporting the funding and independence of scientists to an assertion of the power of the scientific approach to affect 'the general welfare and good of the nation' (Turner 1980: 608; see also Collins 1981: 212, 231).

No existing account quite catches the energy and diversity of Huxley's activity and public presence.[17] It might seem that no association or network in this period was complete without him. He was a key figure not just in the Eugenics Society but also in both PEP and NFYG (see Jones 1986: 113–36). More recently, Peder Anker has provocatively suggested that the planning discourse of the period that Huxley championed derived from ecological styles of thought, citing the significance of the ecologist Max Nicholson's National Plan for Britain, published in the *Weekend Review* in February 1931, which prompted the foundation of PEP (Anker 2001: 208–9). Huxley's interests in Modernist architecture can be seen in the same light (see Anker 2005: 231–2, 239–40). This may be too reductive about the sources and contemporary meanings of planning discourse, but Anker certainly does help us understand the ways in which planning was an essentially scientistic approach to social and economic problems. This extended beyond the economic planning advocated by many left and liberal economists impressed by the Soviet experiment to the application of rationality more generally in what can be seen as a variety of modernism. In this context, it is not surprising to find Huxley quoted in the *Architectural Review* that 'one of the weaknesses of the last thirty-three and a third years has been the lack of planning, which must be the basis of all development from now on' (quoted in Harrison 1934: 190). In this example, we find him responding to the journal's specific concerns and revealing a type of technological enthusiasm like that described in the previous chapter: 'Planning should be nationally financed and regionally controlled ... Airports in the centre of cities will soon have a large part to play; and provision will have to allow for the landing of autogiros' (quoted in ibid.). For Huxley, as for many who became involved in discussions on the social relations of science, it was natural to extend the scientific approach from the field or the laboratory to questions of how human lives should be lived and governed. In this sense, the response of the liberal Huxley to communist rule in Russia is perfectly congruent:

> Science is an essential part of the Russian plan. Marxist philosophy is largely based upon natural science ... Not only does it assert that the method of science is the only method in the long run for bringing phenomena under our control, not only does it assert that this is applicable to social as well as to biological and physical phenomena, but it asserts that the scien-

> tific attitude must form part of the Communists' general outlook. (Quoted in Werskey 1988: 240)

Huxley had been involved with non-fiction films from at least 1929 when he was sent to East Africa by the Colonial Office Committee on Native Education to investigate school biology teaching there. The EMB, which fell under the Colonial Office, was also 'interested to know more about the value of the cinema for education and propaganda purposes' (Huxley 1931: 57) and so equipped Huxley with three different types of film to gauge the responses of African children; *Black Cotton* (from the Empire series, 1927) and two *Secrets of Nature* films: *Fathoms Deep Beneath the Sea* (1922), shot by H. M. Lomas in a Plymouth aquarium, and Percy Smith's *The Life of a Plant* (1926) on the life cycle and growth of a nasturtium, with stop-motion sequences. What is interesting here is to see Huxley's engagement with the possible detailed impacts of different types of films. The cotton film had been chosen because it was expected to be familiar, the sea life film was there to confront the children with unfamiliar content, and the nasturtium film because it was believed that the stop-motion content would make particular intellectual demands on the children. He concluded that 'the cinema could be a most important instrument for awakening the young African mind' (1931: 60).[18]

Other examples of his involvement with films include his participation during 1929, as Chairman of the Association of Scientific Workers, in setting up the Commission on Educational and Cultural Films whose report *The Film in National Life* is generally credited with leading to the foundation of the British Film Institute (see Low 1979a: 182–7). Chairing the first meeting, in April 1929, he set out the rationale, which echoed his African project, 'that films were not being used for pedagogic purposes as much as warranted by their advantages'; his colleague Vernon Clancey made explicit that the brief extended to adult audiences.[19] From 1933 to at least 1937, Huxley was also biological films adviser to Gaumont-British Instructional. His fellow adviser H. R. Hewer explained how it worked:

> [In 1933] a general scheme covering a series of films was made. The biologist decided on the theme of each film and wrote the 'argument', with suggestions on suitable shots, with the advice of the camera-man [Percy Smith is named] and possibly the editor. The first attempt at shooting was then made, sequences built up by the editor, with the biologist as supervisor, until the whole was complete, revised shots being made when necessary. Then the commentary was made, based on the argument, but pruned and refined to fit the final version. *The biologist remained in control throughout,* assisted by the technical advice of the film experts. (1946: 17; emphasis added)

In 1934, Huxley made his first film, the 10-minute *The Private Life of the Gannets*, whose title echoed *The Private Life of Henry VIII* (1933), the previous success of the film producer Alexander Korda, who financed Huxley's project and provided him with Osmond Borrodaile as cinematographer. Huxley also called in his friend John Grierson to advise on structure and to film the final sequences. The film follows the life cycle of the birds from courtship, via hatching and feeding of the young, to the fledgling's first flight. The final sections of the film show the birds catching fish. The commentary concludes in ecological strains reflecting on the birds' likelihood of maintaining their 'present secure position'. It has many similarities with the *Secrets of Nature* films; there are time-lapse sequences of hatching and slow-motion shots of birds in flight. The narrative has the familiar life-cycle structure, although it is without the facetious humour that characterised the G-B I films.[20] With its backing from Korda, it received general theatrical release (see Mitman 1999: 76-9) and received an Academy Award for best short subject in 1937 (see Low 1979b: 121).

In a review in the same year of Mary Field and Percy Smith's *Secrets of Nature* book, Huxley reflected on the 'vastness and importance' of the field opened up by interest films. He commented that 'the subjects are there, the technique is already there ... the public which appreciates such films is there, albeit somewhat scattered'. He perceived two problems: how to render production and distribution more flexible; and how to go beyond existing experience 'for the further extension of the nature film and its relatives into non-commercial spheres – education, general and special research, applied science, propaganda and persuasion and the like' (1934b: 121). Huxley's commitment to factual film was too extensive to explain away as merely being helpful to friends. This is shown by the fact that his contribution to ARFP was not limited to being adviser; he also lectured on their behalf in their courses 'about the social, aesthetic and educational aspects of cinema'.[21]

From the mid-1930s, however, his concern with the social role of science led to him making films in a new convention, stressing this social role. The first of these to be started in or before October 1935, although it was not completed and finally released until 1938, was a film for the Eugenics Society, generally known by the title *From Generation to Generation*.[22] Huxley, who had been supervising the film's production from the beginning (presumably in a joint role as GB-I adviser and active Eugenics Society member) agreed to narrate the film in spring 1937 when the Society's president Lord Horder declined.[23] This was a propaganda film made by Gaumont-British Instructional on behalf of the British Eugenics Society. Although support for eugenics had been widespread amongst liberal and left intellectuals earlier in the century, Huxley was in a minority amongst his peers in retaining his support and continuing his membership of the Society after the rise of Hitler (see Searle 1979: 163–6).

From Generation to Generation was, and was recognised to be, a propaganda film, intended to persuade audiences of key Eugenics Society policies. But it was of an unusual kind because it did not use the dominant 'moral tale' fictional approach generally favoured for health propaganda (see Boon 2005), but adopted an instructional style instead. There is an ARFP advertising leaflet in the film's production files, which implies that the Society was aware of the documentary alternative, although the film had been under discussion two months before the foundation of the Association.[24] The implication of the propaganda was different, however, from the documentary approach. Whereas documentary rhetoric was most often centred on the role of states, corporations and experts, here persuasion centred on gaining support for a single social sanction – the control of fertility.

There were several versions of the film; the title *From Generation to Generation* was used to describe the composite of the biological film *Heredity in Animals* and the specifically eugenic film, *Heredity in Man* (both nominally 1937), which is the major concern of this discussion. The first of these, drawing on GB-I's extensive experience with *Secrets of Life*, showed the physiology of reproduction, illustrated by moving diagrams and microcinematography, followed by sequences of Mendelian genetics in dogs, horses and other animals. *Heredity in Man* was directed by J. V. Durden, a staff member at GB-I, under the supervision of Huxley and Hewer. It is comprised of human pedigree charts with live-action sequences, illustrating the inheritance of physique in the case of the Phelps family, oarsmen for generations, and of mental 'defect' in the case of the Mason family. Next, a sequence shows 'defectives' at Stoke Park Colony (an asylum near Bristol, whose superintendent was an active member of the Society) engaged in occupational therapy. We then return to the pedigrees: of musical talent in the Godfrey family; and of theatrical abilities in the Terry-Gielguds, including the actor John and the BBC radio producer Val. The film ends with a chart showing differential fertility between 'dysgenic' and 'eugenic' classes with Huxley voicing the conclusion: 'If we are to maintain the race at a high level mentally and physically, everybody sound in body and mind should marry and have enough children to perpetuate their stock and carry on the race.'[25]

The legalisation of voluntary sterilisation of the 'mentally deficient' was the major policy aim of the Society in the 1930s (see Jones 1986: 88–104). But the nearest the film gets to stating this directly is where Huxley's commentary asserts that 'once they have been born, defectives are happier and more useful in these institutions than when at large. But it would have been better by far, for them and the rest of the community, if they had never been born.'[26] The film's commentary is diffident about its main propaganda points, but at the same time there was an almost incongruous investment in the power of the image. It is as though the Society expected the main work of the film to be done by the charts, and for the live-action sequences to serve as emphasis. But the sequences of the 'defectives' were also in some sense the point of the film; the propaganda secretary implied as much when she reflected later that 'the general public has little or no idea of what mental defect means and some of the effects of this part of the film are lost because it is shown too quickly'.[27] In particular, the Society invested heavily in the power of close-ups of members of the Mason family. So powerful did they consider such representations to be that they investigated legal means to protect themselves against possible libel suits brought by the Mason family (see Boon 1990). The *coup de grâce* finally came in January 1938 when *Heredity in Man* was submitted to the censors. A note records that

> they did not pass it for public exhibition as they took exception to the defective family. It could therefore only be used for private shows where no money was taken. In April, it was again submitted to the Board with the imbecile family removed. It was then passed for public exhibition with an 'A' certificate.[28]

The propaganda secretary commented in the *Eugenics Review* 'only the pedigrees and the commentary remain. These cuts do seriously mutilate the film' (Anon. 1938d: 203–4).

The Society's relations with Hogben and Haldane were stormy throughout the 1930s, and although Huxley stayed within the fold he was not averse to using the Society's platform to express critical changes of opinion, notably – following an intervention from Hogben (see Werskey 1988: 241) – about the relative significance of heredity and environment; in his Galton Lecture of 1936 he said:

> [Eugenists] find themselves in apparent conflict with the environmentalists and the protagonists of social reform. Speaking broadly, the field of human improvement is a battlefield between eugenists and sociologists … We eugenists must no longer think of the Social Environment only in its possible dysgenic or non eugenic effects, but must study it as an indispensable ally. (1936a: 31)[29]

Appreciating this change of opinion allows us to see how Huxley could, with consistency, be involved in both the nutrition and the eugenics film. There are signs that his attitude to the eugenics film altered in the light of his changed priorities, as a document modified by him directly addresses its genetic stress:

> No attempt is made in the film to distinguish between the effects of heredity on the one hand and on the other of such environmental influences as the family tradition and training. The observer is left to infer that the outstanding abilities here illustrated would not have manifested themselves so brilliantly unless there had been a marked hereditary and inborn predisposition, upon which the educative forces of a favourable environment could exert its moulding influences.[30]

The Society was sensitive to the accusation that it sought to mislead on such matters. Just before shooting began, the Society's General Secretary C. P. Blacker had written:

> It would be undesirable for us to lead the public to believe that good oarsmanship, mental deficiency, histrionic talent etc. conform to principles of heredity which we understand as clearly as we understand the transmission of tallness in peas ... In my opinion, the Eugenics Society has, in the past, been justly criticised for suggesting to the general public that our knowledge of human abnormalities is much greater than it is.[31]

This is typical of the manoeuvres that Blacker made as he sought to retain support for the Society. The film may be seen as a transitional project for Huxley. His involvement with biological films continued but, with the foundation of ARFP, he also became active in the documentary mode, expressed within *Enough to Eat?*, the main focus of this chapter. But, by the standards of Huxley's associates concerned with the social relations of science, the eugenics film was reactionary, and by the standards of those in documentary, it was also stylistically plain.

The scientific context of *Enough to Eat?*

Although *Enough to Eat?* was produced at the same time as *Heredity in Man*, and although the two films have many superficial similarities, not least the presence of Huxley as 'front man', their representational styles and their arguments are different, both in terms of content and political implication. *Enough to Eat?* was concerned with nutrition science, but this subject was far from uncontroversial in 1936. To understand its sensitivity it is necessary to grasp the way in which it had been politicised earlier in the decade. It had already been substantially aired in public, notably with the clash over the minimum dietary standards (and the costs entailed) proposed by the Ministry of Health's Advisory Committee on Nutrition and those suggested by the BMA's Nutrition Committee (see Smith 1986: chapter 3). Ministry of Health officials resisted the conclusions of scientists such as Orr because of the financial implications of increased provision of money or intervention foods to those on low incomes (see Smith 1986; Mayhew 1988).[32] In addition to this economic argument, there was a cultural factor relating to the view of nutrition held by the Ministry. It was characteristic of officials – who included many medically qualified specialists – and Ministers to resist alternative sources of expertise to their own. But we should not necessarily assume these to be inappropriate views for them to hold. For a properly symmetrical account, we must see the champions of the 'newer knowledge of nutrition', in alliance with those responsible for social surveys of malnutrition, as an interest group proposing a new vision of public health based on recent science, and thereby contesting the Ministry's view of the appropriate means of achieving improvements in the population's health. Edward Mellanby, for example, in 1927 had portrayed the Ministry's Food Section as perpetuating a hygiene-based notion of food and health, accusing them of wasting time 'chasing adulteration, food poisoning and milk grading' at the expense of 'sufficient public education in health' (quoted in Mayhew 1988: 447). The contestation of public health policy by nutrition enthusiasts may also be seen as an example of the 'outsider politics' of scientists in this period, proposed by Gary Werskey (1971) as a way of describing scientists' attempts at gaining political influence. And, in a period when the state was extending its reach, it may have been reasonable for scientists to assume that there was everything to play for in asserting the virtue of a state expanded to take account of scientific opinion.

Orr became a key advocate of nutrition science after Mellanby, who had become Secretary of the Medical Research Council in 1933, 'became conspicuously uninvolved in the increasingly political "Nutrition Movement"' (Smith 1986: 172). Orr had spent the interwar years building up the Rowett Institute near Aberdeen. This he achieved using techniques of network building which would be familiar to any disciple of Bruno Latour. For example, his association with Walter Elliot – a friend since they were undergraduates at Glasgow – was very productive. Elliot was the Conservative MP for Lanarkshire and a powerful figure at the Empire Marketing Board, who until about 1926 had spent parliamentary recesses at the Rowett undertaking research on animal diets and the mineral content of pastures.[33]

Orr's *Food, Health and Income: Report on a Survey of Diet in Relation to Income*, published in March 1936, concluded that less than fifty per cent of the population ate a properly nutritious diet. Despite its semi-official nature – Elliot, as Minister of Agriculture, was responsible for the department which commissioned it – he had been prevented from publishing it except as his personal work, the Cabinet having refused to issue it as an official publication, even with a special preface (see Branson & Heinemann 1973: 232). Orr, however, did delay six months beyond the 1935 election – which returned a second National government – so as to avoid 'doing any damage' to the Conservative Party (see Smith 2000: 64–73). In the event, it was published by Harold MacMillan, Conservative MP for Stockton.[34] MacMillan could have been expected to be particularly sensitive to these issues, as Stockton's Medical Officer of Health, George M'Gonigle, published in the same year *Poverty and Public Health*, his analysis of the effect of new housing costs on the nutritional status of ex-slum dwellers. Orr himself drew the explicitly political moral of *Food, Health and Income*:

> If these findings are sufficiently accurate to form a working hypothesis, they raise important economic and political problems ... One of the main difficulties in dealing with these problems is that they are not within the sphere of a single Department of State. This new knowledge of nutrition, which shows that there can be an enormous improvement in the health and physique of the nation, coming at the same time as the greatly improved powers of producing food, has created an entirely new situation which demands economic statesmanship. (1936: 50)

The impact of the Depression on the nutritional status of the unemployed had already received more explicitly political treatment by the Communist Party-affiliated Kino in their *Hunger March* newsreel and the drama *Bread* (both 1934) (see Hogenkamp 1986: 110–13). Like *Enough to Eat?*, these films brought malnutrition to public attention, but their solutions were political, not scientific.

Huxley had a core role in *Enough to Eat?*. He spoke the commentary, appearing seated at his desk at London Zoo. This is one way in which this film, which to our eyes may look rather conventional, was revolutionary. Earlier films had used journalists to speak commentaries – *The Face of Britain*, for example, used A. J. Cummings of the *News Chronicle* – but they were never seen in vision. The only previous occasions on which British scientists had appeared on screen was in the Gaumont-British Instructional series *Eminent Scientists* (1934), made for the National Physical Laboratory, which featured eight half-reels of different scientists speaking to camera, including William Bragg, Lord Rutherford and Oliver Lodge, in brief, lecture-style reminiscences of their scientific research (see Low 1979a: 27–8, 215). But in *Enough to Eat?* Huxley was doing two new things: he was appearing as the film's 'an-

Fig 3.1 *Enough to Eat?*: Julian Huxley, public scientist

chor' or 'presenter', the personality carrying the film. But at the same time, he was doing this *as a scientist.* Furthermore, he was using this novel position to make a public argument about malnutrition and by that very action asserting not only the appropriateness of a scientist making this quasi-moral and political argument but also the relevance of science as the appropriate means to diagnose and cure social ills. In this, the film added an explicit additional implied claim about the beneficial impact of science on society to the slightly older model, which emphasised the benefits of technology as applied science or the concurrent model that stressed industrial technology as the historical route to modernity.

Given that 'public science' can be dated to at least 1880 (see Turner 1980), the apologists of professional science had not been quick to turn to cinema. As we have seen in previous chapters, for most of cinema's first four decades where science was present in films it was as lay natural history. There was a small admixture of 'interest' films showing technology *qua* industry, but it was only in the mid-1930s that such representations became resolutely modernistic and began to concentrate on the higher technologies of telecommunications and electricity. Only with *Enough to Eat?* do we see the emergence of a novel form of scientific film, which projected the scientist as an agent of moral concern. In asking why the exponents of public science did not use film earlier it is important to reiterate the factors we have been discussing that provided the essential 'conditions of possibility': the rise, with the documentaries of the British group, of a mode of cinema that joined public relations, citizenship discourse and visual representation; the circumstances of the Depression, which invaded both the public conscience of science and the practice of cinema; and the associative culture that provided opportunities for the cross-fertilisation of different aspects of science, filmmaking and politics.

Huxley may well have been involved in films before, but he is not normally associated with nutrition science, and might not seem the obvious person to choose as the main voice of a film on this subject. Yet, in the eight years' gap of formal employment between King's College London in 1927 and London Zoo from 1935, nutrition was one small aspect of his polyvalent activity. A crucial source for this was his work with PEP, as chair of its research group. In this context, nutrition became the paradigmatic example of the role of scientific research in relation to society. Research was a key activity for PEP, and Huxley's committee became the heart of this. His group had started with a joint brief in research and statistics, but had split into two separate sections. Huxley explained the work of the Research Committee in 1933. He reported that 'the financing of research was unorganised, and the public interest should be awakened in research as a national effort. ... He mentioned the difficulty of getting the results of research passed on to the consumer in the shape of reduced costs'.[35] Its work continued throughout the decade.

In autumn 1934, his committee produced two 'broadsheets' for the PEP journal *Planning*. One was concerned with 'consumer research' and the other, an exemplification of the former, with the need for a national food policy. The logic of the argument flowed from research to the consumer; the report asserted that 'specialists qualified to speak are now able to promise rapid results if only it were possible to translate scientific knowledge into daily practice in the nation's homes' [sic].[36] The issue became one of how to create demand for the forms of research most valuable to society. The title of a later broadsheet, 'when will consumers wake up?', sums up their view. 'So little specific research and thinking has been done on behalf of the consumer in the United Kingdom' it asserted, 'that even the most elementary information on the subject has to be gathered in fragments from many sources' (Anon.1936k: 8). In the national food policy broadsheet, the argument was based on a certain definition of consumer research:

> The task that confronts us therefore is not merely to satisfy hunger where it still exists, to liquidate scarcity and put plenty in its place, but so to guide the choice of the consumer that we shall as a nation choose so far as possible the best type of plenty for the health of our generation and for the future of the race. (Anon. 1935: 2)[37]

Gaining knowledge of the consumer in relation to diet was precisely what Orr had been doing in his surveys of income and food expenditure. He was a correspondent and interviewee of the Research Committee as it considered food policy.[38] They linked consumer research to science, and together they make the case for planning:

> We are just acquiring a body of knowledge which will, if it is sufficiently widely spread and used, help us to choose on a basis of science and informed judgement. But this means planning. The waste and misery due to the failure to apply knowledge which we already possess is obvious in every British town and village. Planning can give a clear and convincing answer to this challenge of poverty and ignorance. The necessary adjustments are too large and complex, and too interdependent, to be brought about haphazard. (Ibid.)

It is worth noting that several people involved in developing the broadsheet were later involved – centrally or peripherally – with making *Enough to Eat?*, including Huxley, Orr, Frederick Gowland Hopkins, Edward Mellanby and Clement Leslie, the gas industry public relations officer.[39]

Making *Enough to Eat?*

Enough to Eat? was directed by Edgar Anstey, who had worked at the Building Research Centre of the Department of Scientific Research before joining Grierson at the EMB in 1931 (see Orbanz 1977: 90). He later recalled that it was Ritchie Calder who persuaded them to make the film (Anon. 1982: 12). In 1934 Anstey had worked as a producer for the Shell Unit and in 1935 he co-directed (with Arthur Elton) *Housing Problems*, the first of the gas-sponsored films. In the midst of making the nutrition film, his employment at the British office of *The March of Time* newsreel was announced, suggesting that work on the film may have been compressed by competing obligations (see Anon. 1936g). The film titles feature prominently the initials 'ARFP' after Anstey's credit, explicitly locating it for us as one of the new non-governmental films. Accounts of the production process state both that it was the industry which chose the subject (see De Mouilpied 1937a: 408), and that 'early in 1935, the British Commercial Gas Association commissioned Elton and Anstey to produce a programme of five films, *inviting their collaboration in the choice of subjects*' (Arts Enquiry 1947: 53; emphasis added). If we take it that those concerned were already familiar to each other as members of an extended network with shared interests and ideological commitments, these accounts are not incompatible. Many of the various scientists and politicians who agreed to appear in the film, or to give advice on its contents, were already associated, either by long-term friendships (as in the case of Orr and Elliot) or by involvement in campaigning voluntary associations, such as the Committee Against Malnutrition, with which Hopkins, Haldane, Huxley and Orr were all associated (see Smith 1986: 169; Anon.1934b: 1358–60).

The formation of ARFP was crucial to the choice of subject of *Enough to Eat?* For its three scientific advisors, as for other prominent biologists, the 'newer knowledge of nutrition' offered an alternative to established modes of argument over the public health, which ranged from older hygiene-based approaches to a then more recent stress on personal responsibility for conduct and health (see Armstrong 1983). Hogben, for example, in his review of Rotha's book, specifically stated that

> Money and effort spent in diffusing political propaganda might be far more usefully employed in promoting *ad hoc* societies to finance the production of documentary films dealing with specific social issues of national importance. Thus the widespread under-nutrition that results from unemployment has prompted protests from Sir Gowland Hopkins in the presidential chair of the Royal Society and Sir John Orr at the British Association. (1936a: 8)

Where older dietary theory rested on the calories consumed by individuals in various occupations, the 'newer knowledge' stressed the role of the recently discovered biochemical micronutrients, that is, vitamins and minerals (see Smith & Nicolson 1989). Advancing in public this specific style of argument allowed these scientists simultaneously to assert the general relevance of scientific research to national policy. Views of nutrition science based on the 'newer knowledge' were widely held amongst those scientists who advocated the virtues of the medical sciences as a specific approach to medicine.[40]

Enough to Eat? falls into five substantial sections: an introduction; laboratory and social survey research in nutrition; the activities of local, national and international organisations; proposed policy for England; and a conclusion.[41] The film's soundtrack is continuous and mainly comprised of Huxley's narration, with no music or other sounds. Its visuals have been edited to fit the soundtrack, and are of several kinds: literal, whether synchronised as in the case of Huxley's appearances, or tied to the soundtrack, as when Huxley states 'here you see the boys of the school being weighed and measured', or literal and specific, as in the use of animated diagrams. There is also substantial use of figurative visuals, as in the montage of publications at the start, and the use of suggestive shots to accompany abstract statements on the soundtrack, as for example where the pitch forking of hay is used to accompany a speech from Lord Astor on world trade. The film's introduction, spoken by Huxley commences:

> Many people in this country are not properly fed. Scientists for years have been studying the ways in which food affects health and growth and the

results of a great deal of their work were made known early in 1936. Their conclusions were widely reported in the newspapers and they shocked the public conscience. The scientists said that the health of the nation was being gravely affected by insufficient food and wrong feeding.

A mixed montage of reports, books and newspaper coverage visually accompanies this speech, which in the second sentence of the film asserts the social concern of scientists. This was intended to state, as Anstey explained to Mellanby, 'in terms of topical events widely reported in the press ... to bring the subject alive by giving it contact in a dramatic way with the daily life of every member of the audience'.[42]

The first main section presents the work of scientists in nutrition science:

> In 1912, Sir Gowland Hopkins, late president of the Royal Society, whom you see here in his Cambridge laboratory, drew the attention of the scientific world to the existence of vitamins, those organic constituents of common foods that are needed for proper growth and to give protection against disease. In 1923, Sir John Orr, who is director of the Rowett Institute and of the Imperial Bureau of Animal Nutrition proved the value of the traces of minerals present in foods. In 1932, Sir Robert McCarrison's work in India showed that the great variation in physique between different Indian races was a direct result of the differences between their diets; those southern races that live principally upon rice are poor in physique, as compared with the northern races which eat milk, fruit, vegetables and meat.

Next, Huxley introduces the milk-feeding experiment of Dr G. E. Friend, Medical Officer at Christ's Hospital Blue Coats School, which acts as an introduction to 'this question of how food consumption varies between [social] class and class'. Huxley concludes:

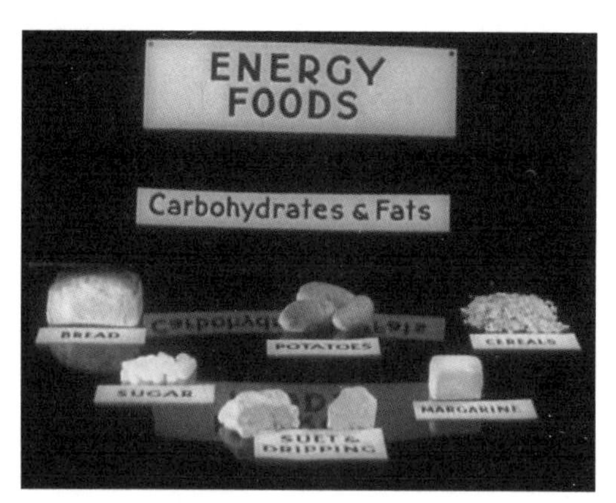

Fig 3.2 *Enough to Eat?*: teaching nutrition

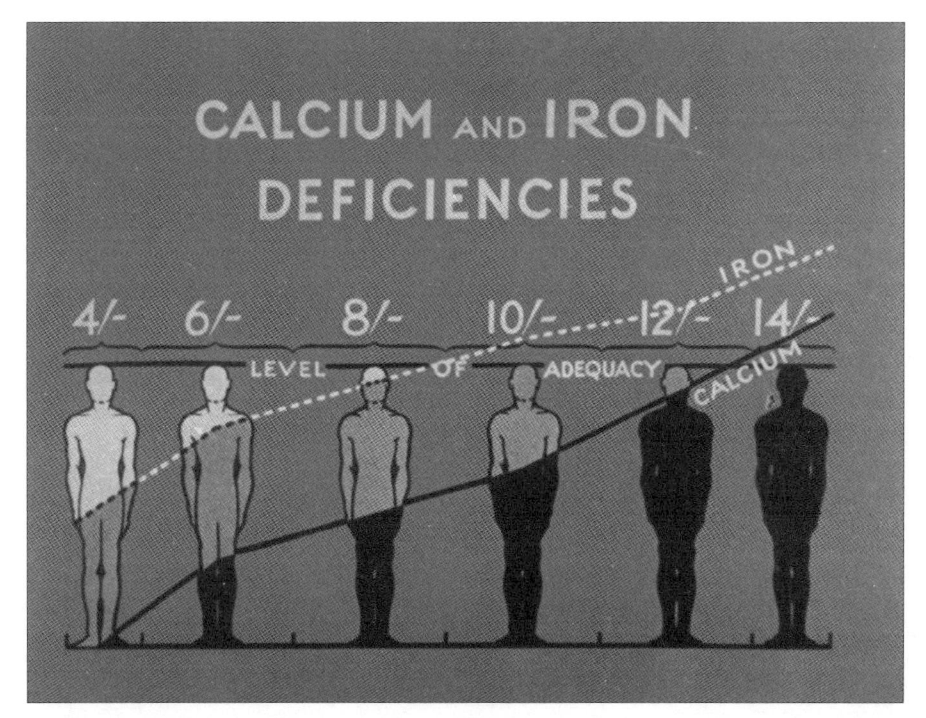

Fig 3.3 *Enough to Eat?*: nutritional deficiency diagram

> Perhaps the most important point established was that the addition of a pint of milk to a quite generous diet made boys grow taller and heavier, reduced their proneness to various diseases and gave them more physical and mental vitality. There is a marked difference in the heights of boys drawn from different classes of society.

The second section, commencing with an introduction of nutrient types – classified into energy foods, proteins and protective foodstuffs – and their sources in everyday foods, proceeds to give an exposition of the analysis and findings of Orr's *Food, Health and Income*, mainly through diagrams that map first the consumption of milk and bread within six income groups, then calcium and iron deficiencies in the same groups. Huxley sums up: 'Sir John Orr's conclusion is that a diet really adequate for health is eaten by barely half the community, whilst that of the poorest class, numbering at least four and a half million people, is deficient in every constituent.' This is followed by a camera interview with the Medical Officers of Health, Vynne Borland (Bethnal Green) and George M'Gonigle (Stockton) supported by two restaged social survey interviews with 'working-class wives'; Mrs Appleby with 2/8d per family member for food per week, and Mrs Sivewright

Fig 3.4 *Enough to Eat?*: Mrs Sivewright, 'working-class wife', interviewed about her diet

with 4/-. The latter, for example, states 'I've seen my husband and myself go rather short sometimes to give the children a dinner. Towards the end of the week of course, we've had bread and jam or bread and margarine and a cup of tea.' In formal terms, this reprise from *Housing Problems* of working-class women speaking within a film, is the second major event in the film – after the diagrams – and the first sequence in which there is a break in Huxley's narration.

At the start of the third section, Huxley opines:

> I am the secretary of the Zoological Society, and at the Zoo we have found the absolute necessity of giving the animals proper food. If we gave the gorillas and chimpanzees the type of diet actually eaten by millions of human beings in this country, we should be rightly blamed for not keeping them in proper condition.

There is then a rather throwaway reference to a planned food policy: 'Nutrition is bound to play an important part in the national policy of this country. Many government departments are directly concerned with it. In fact we really need a national food policy.'[43] He then turns to look at 'what is actually being done about this national problem', starting with the level of government policy, showing the work of the Milk Marketing Board, including what looks like a newsreel sequence of Walter Elliot at a school launching the cheap school milk scheme in 1934. The Shoreditch Maternity and Child Welfare Centre is given as an example of a part-nationally-funded institution providing food for 'mothers with small incomes'. The London County Council, in a speech by Herbert Morrison, its leader, is given as an example of local authorities' work in nutrition. Finally, in this section, Viscount Astor speaks on behalf of the League of Nations Nutrition Committee, linking poverty, ignorance and agricultural production:

> Ignorance about proper diets is widespread and poverty is the main factor in malnutrition. Governments must face this and take proper steps to deal with it … Most countries also will have to import considerable quantities of food if the cost of living for the poor is to be kept low. This means that a wise nutrition policy will help to restore world trade and so be an important factor in the preservation of peace.

Huxley starts the fourth section with the words 'and now; what must we do in England?', followed by M'Gonigle's assertion that 'the average working-class housewife, by rule-of-thumb methods, knows pretty well what foodstuffs to buy to feed her family properly, though she doesn't know the difference between a vitamin and a bus-ticket'. This is a statement of one side of the 'poverty versus ignorance' argument, which has been discussed by David Smith and Malcolm Nicolson. Broadly, at this stage in the argument, most advocates of the 'newer knowledge of nutrition' maintained that poverty was the principal cause of malnutrition, arguing that the more nutritious foodstuffs were not afforded by those on low incomes. The Ministry of Health stuck to the older calorie-based approach of the 'Glasgow School' of nutrition, whose champions contended that lay ignorance of sound nutrition was the problem and that instruction of the populace would therefore ensure adequate nutrition levels. Indeed, the Minister Kingsley Wood used precisely this argument in a House of Commons debate in July 1936 (see Smith & Nicolson 1993: 223 and 1995: 288–318). Huxley sums up the film, suggesting radical solutions in terms of policy: 'No complete solution of our problem is possible without considerable economic changes, either by providing the lowest-paid members of the community with increased purchasing power or with cheap or free milk and other protective foods.' But the impact of this statement is immediately reduced by the sentences that follow, which lay stress on the instructional solution to the malnutrition problem: 'But such a solution is a difficult long-term matter which will need all the community's patience and ingenuity. Meantime, for a large number of people, particularly in the higher income classes, much good can be done within the present limitations by teaching the proper choice and use of foods.' The film returns to its classification of food types, recommending a diet for a working man and a child. Against a montage of scenes from earlier in the film, Huxley narrates eight closing sentences, making a concluding link between science and living standards, finishing with shots of the maternity and child welfare centre and the two 'working-class wives':

> The conscience of the nation is aroused. The new scientific knowledge has shown us a new standard of life at which to aim. A new charter is being

> written for the people of the country.[44] By whatever means the problem is tackled, a better level of health and nutrition must and will be achieved.

Superficially triumphant, this conclusion proposes no duty of policy formation to government, but in the phrase 'the conscience of the nation is aroused' contains a nascent citizenship motif: the final sentence is all in the passive voice, implying that 'the nation' will 'tackle' 'the problem'. The earlier script had been much harder hitting in its conclusion:

> The statesman and the scientist have to face up to the problem of removing the poverty handicap and the ignorance handicap which are preventing the housewife from getting rid of much of the country's ill-health and setting a new standard in national well-being.[45]

Elizabeth Coxhead, reviewing the film in the *Left Review* stated: 'Mr Anstey, never straining after mere cleverness, uses the cinema's power of illustration and contrast merely to bring home his facts' (1936: 777). Anstey later described *Enough to Eat?* as 'a scientific argument deployed by scientists'.[46] At the time, he stressed the power of the scientist's narrative; writing to Mellanby about the draft script's reference to the Medical Research Council, he stated: 'I believe this can be done most convincingly by introducing a well-known research worker who will talk from personal experience.'[47] All this raises the question of cinematic form. Within Rotha's classification, *Enough to Eat?* was a 'descriptive or *reportage* film', or in Grierson's 1935 term, an 'information film' (quoted in Rotha 1973: 106). If we accept Rachael Low's judgement that Anstey could employ an editing style 'like a more sophisticated and modern application of the Russian style used in *Drifters*' for the 1933 film *Granton Trawler* (1979a: 60), it follows that the adoption of the *reportage* style was a deliberate choice to represent a social and scientific subject in the supposedly more neutral and 'scientific' style of a film lecture. This choice may have been compounded by the controversy surrounding the nutrition question in 1936; those involved had made a deliberate choice to intervene in a politically disputed area, and the adoption of an 'objective' form was a rhetorically sophisticated move.

We may briefly consider *Enough to Eat?* in terms of associative culture if we think of the network of people involved in its production as what Hogben had described in his book review as '*ad hoc* societies to finance the production of documentary films dealing with specific social issues of national importance' (1936a: 8). The first point to note is the very wide range of political sensibilities represented amongst the participants, from the Labour politician Herbert Morrison, to Lord Waldorf Astor, an American-born British businessman (publisher of *The Observer*) and Conservative politician.[48] M'Gonigle occupied more ambiguous territory. Although his politics are far from clear, he was regarded by senior figures within the Ministry of Health as 'a promising Labour politician'.[49] However, although the argument of his book *Poverty and Public Health* was welcomed by figures on the left, as we have seen, it was also accepted by some on the right such as Astor.

The performances of a range of scientists present in *Enough to Eat?*, of whom Huxley is only the most prominent, are also significant as public science. Gowland Hopkins made an appearance at the beginning (although it is likely that this footage was re-used from the Cambridge University student film *Vitamins* (1936)). It is Orr who is otherwise the most visible scientist there. He is mentioned in the list of three researchers into the 'newer knowledge of nutrition' at the start of the film, and shown on the Rowett Institute's farm. Animals from his experiment in feeding different rats 'a typical poor working-class diet' and 'a good diet' are shown. *Food, Health and Income* dominates the statement of the film's argument. Calcium and iron are used to represent the adequacy of nutrition, leading the balance of the film to favour Orr's work on minerals rather than other researchers' work that stressed vitamins within the 'newer knowledge of nutrition'.

This emphasis on Orr is no coincidence, given the prominence of *Food, Health and Income* in the national debate. He was an adept public scientist, who had seen the potential of documentary for furthering his interests as early as 1929, when he was involved in a projected film for the EMB, entitled *Grasslands of the Empire*, which was later cancelled.[50] This grasslands film established relationships which were later to bear fruit, not only in *Enough to Eat?*, but also in the 1931 film *An Experiment in the Welsh Hills*.[51]

In preparations for *Enough to Eat?*, not only did the film-makers go to Aberdeen to film at the Rowett Institute, but Orr went beyond agreeing to appear, contributing suggestions about how to dramatise part of the scenario. He suggested that Mellanby (Secretary of the MRC), Cathcart (Regius Professor of Physiology at Glasgow University) and he, 're-enact for the film a meeting between the British members of the Technical Committee of the Health Section of the League of Nations'.[52] Mellanby's reply to Anstey, coming just a month after the publication of *Food, Health and Income* had occasioned a seven-and-a-half-hour debate in the Commons (see Mayhew 1988: 458), gave a telling indication of the public scientific manoeuvres of an individual at the heart of the scientific establishment compared to Orr at the periphery:

> It would be impossible for me to agree to this suggestion. In the first place, my position as a government servant and as a medical man would prevent

> my doing this, and in the second place there is absolutely no justification for suggesting that Sir John Orr, Professor Cathcart and myself ever discussed the plan of action to be adopted before the meeting of the Technical Commission held in London in November 1935.[53]

We may note that for Mellanby, factual inaccuracy came second to his sense of official propriety. But, by virtue of being Head of the MRC, Mellanby had become a government insider. Orr continued to be, to a degree, an outsider, contesting public health policy without the means to effect any change directly, despite his closeness to Elliot. This 'outsider' attitude was shared by the documentarists, who viewed their activities as a nascent state information service whilst holding views which very often went well beyond government policies. The implication is that, just as they considered that documentary film should have a larger role in the life of the state, they also contended that government should take a more active role in the alleviation of malnutrition. For documentarists, *Enough to Eat?* was the most outwardly political film they had made. Certainly, neither of the 1935 social reportage films *Workers and Jobs*, made for the Ministry of Labour, nor *Housing Problems* presented arguments about subjects that were politically inconvenient for the government. But the documentarists relished their advocacy of the sensitive topic of nutrition. Under the headline 'Whitehall defaults', their journal *World Film News* reported:

> There are two curious omissions in the nutrition film. The film talks of a food policy for the nation, but the word is spoken by Sir John Orr [actually Huxley] and not by the Minister of Agriculture [Elliot up to 28 October]. It describes one of the principal issues before the Minister of Health, but there is no word from Sir Kingsley Wood. Is it possible that the business of national education is passing by default from the offices of Whitehall to the public relations departments of the great corporations? It might seem so. (Anon. 1936h: 29)

In fact, as late as June, Anstey had expected that both Ministers might appear,[54] and in late August a Ministry of Health official stated that a decision had not been made by either Ministry.[55] But, given that *Enough to Eat?* was filmed and edited within three months of the House of Commons debate in which Kingsley Wood had been obliged to defend the government's nutrition policy against opposition attack, it is scarcely surprising that he did not appear (see Anon. 1936m). The same increase in sensitivity of the nutrition question might also explain Elliot's presence in the film only in library footage.

Enough to Eat? deliberately avoided coming down on either side of the 'poverty versus ignorance' arguments so prominent in 1936 as is clear from the evidence of the draft scenario sent to Mellanby in June.[56] In the end, it is the citizenship motif of the film's close, so typical of documentary, which carries the day; the implication is that citizens must solve the problem. Paul Rotha picked this up in a newspaper article: 'Films like the nutrition film, if widely enough shown, must help people to realise that only by united effort can these wrongs of underfeeding be put right' (1936b: n.p.). Overall, we may see *Enough to Eat?* as an important departure for the documentarists; working with new allies they had made an explicitly political film embodying a contestation of the Ministry of Health's public health policies. In addition, they had taken the *reportage* film style and moulded it into a scientific lecture format which was intended to imply a neutrality about the views stated within it. For a documentary, the film was very widely shown; it was seen in specialised circumstances, including a meeting of the Committee Against Malnutrition on 5 November, when Harold MacMillan was in the chair (see Anon. 1936n). But it was also more generally available, having three showings a day for at least two weeks after its release at the Building Centre in London (see Anon. 1936k: 8). The gas industry provided a free showing to anyone with access to a hall and an audience of around 150 people. As a result, it was widely seen outside London, including Grimsby (Anon. 1936o), Newcastle (Anon. 1936p) and Manchester (Anon. 1936q), sometimes with speeches and discussions ensuing. It was also widely reviewed in journals including the *Times*, *Sight and Sound* and many others. But perhaps the most valuable assessment to this account is Julian Huxley's own, recorded in his journal: 'V. interesting – 1st attempt at film lecture. But (a) too long (b) confused and confusing in parts (c) not enough use of diagrams or sharp dramatic contrast (d) not enough pauses in dialogue.'[57] Not only does Huxley's criticism chime with that of one of the few critical reviews – by Arthur Vesselo (1936) in the *Monthly Film Bulletin*, a review in cinematic terms – but it also reveals the degree of Huxley's engagement with the film medium for scientific use.

Huxley and his colleagues at PEP expressed their satisfaction with the way things had gone. An issue of *Planning* in December 1936 was devoted to the malnutrition controversy. Under the heading 'The Scientist Butts In' they opined:

> For those who believe in the possibility of creating a happier world through the bringing of science into working partnership with democracy, it is on the whole satisfying to trace the emergence during the last four years of malnutrition as a public issue ... Owing to the researches of a number of sci-

> entists … and to the bold and vigorous initiative of several men who dared to step outside their laboratories or offices to insist that the public should understand the implications of the new discoveries, malnutrition has become one of the most talked-of public problems of the day. (Anon. 1936r: 1)

It is interesting that, in this context, at the end of a substantial bibliography to the summary article on the controversy, *The Nutrition Film* is listed as a resource (Anon. 1936r : 14).

Nutrition, smoke and architecture food: films and expertise from 1937

Enough to Eat? was not an isolated case. Films, often in the reportage style, that represented planning or scientific expertise in the service of society proliferated in the later 1930s. Nutrition was, as we have seen, not only the paradigmatic case of scientific knowledge within discussions of the social relations of science, but also particularly sensitive political territory for the documentarists to choose. But Huxley and the documentarists kept to this subject, producing several further nutrition films. After the nutrition film, Anstey moved full time to *The March of Time*, where he set about making a story about malnutrition for the newsreel. Huxley had evidently relayed his criticisms of the earlier film to Anstey, who wrote to him in October 1936:

> I hope we shall develop this kind of film into a pretty powerful instrument – with greater attention to the visuals as distinct from the sound. Because in the first attempt as you have seen for yourself I have gone to the extreme of sacrificing many of the demands of the picture in the interests of the speech. Between this extreme and the diametrically opposed style of the early documentary there lies a wide field to be cultivated and I am sure we shall be making use of the kind of devices you suggest to bring our arguments alive.[58]

When the British edition appeared in January 1937 (see Anon. 1937b: 14) under the title *Food or Physical Training?* (second year number seven), it presented an argument between fitness and malnutrition. This had become a matter of public and political debate earlier in 1936, especially after the publication of *Food, Health and Income* in March, the same month that, against the background of the rise of the Fascist powers, newspapers had announced that Britain was to rearm (see Rotha 1973: 164). Starting with a sequence on traditional English fare, it quickly moved to shortages of food for the unemployed, followed by a short speech from Alfred Duff Cooper, the Minister of War, to establish the recruitment part of the story. Over shots of testing of recruits, the commentary intones: 'But, as recruiting starts, England discovers an appalling fact: more than half the applicants are physically unfit for national service.' A recruit explains that, being unemployed, he cannot afford the food that would enable him to put on weight. A caption puts this down to malnutrition. Hunger marchers are seen. The nation is said to be shocked by the findings of *Food, Health and Income* – Orr is seen in a sequence from the earlier film – which is followed by Huxley's speech from *Enough to Eat?* on animal feeding at the Zoo. Tom Johnston, Labour MP, is seen recreating his speech from the House of Commons advocating that agricultural surpluses be redirected to the hungry.[59] A sequence on the Fitness Campaign is concluded with the damning commentary: 'As the physical culture fad sidetracks public attention from malnutrition, still without food are Britain's jobless millions.' The Archbishop of York re-enacts dictating a letter to the *Times* on balancing nutrition with fitness. Meliorising activities are shown, including milk bars and a cookery demonstration by the Gas Light and Coke Company, which features a 'working-class wife' stating the impossibility for the poor of affording the levels of milk consumption proposed by nutrition scientists. There follows a sequence representing the Peckham Health Centre (a unique experiment in social medical surveillance mixed with a membership club; see chapter five) as an institution combining good nutrition and exercise, the model for 'perhaps 2000' to be provided by the State. The commentary says that 'welfare experts' have turned their concern to 'the younger generation' and 'warn that unless that they are somehow cared for by the State, nearly half of all the children of England will continue to be undernourished'. It finishes, over shots of children eating communally, with the commentary: 'Well does Britain know that a healthy people is the best defence for peace at home and abroad.'

The film provoked considerable controversy, with at least three London cinemas refusing to show it, and the powerful Cinema Exhibitors Association opposing its projection on account of its lack of patriotism (see Anon 1937c; Anon. 1937e). At a meeting of the Association, one member 'protested at certain statements made … which might give the impression that monkeys in the Zoo were better fed than the children in Great Britain … he was not going to show such an episode' (see Anon. 1937d: 1).

Huxley's compromise position on nutrition came to the fore in one final 1930s film, *Plan for Living* (1938), directed by Donald Carter for Gaumont-British Instructional. Huxley is clearly the 'star' of this film, as it starts with a shot of his entry in *Who's Who*. Once again visible on camera, he states:

> I want to tell you something about the rules of good feeding. These can be put under three main headings: first, there are the different kinds of work

> that different foods do and why you should eat them. Secondly: however limited your food budget may be, know how to make the best of it by your shopping, and thirdly: know how to cook food well so as to preserve its value and how to serve it in the most appetising way.

Addressed to the imagined housewife viewer, this film reveals its more instructional nature by using direct second-person address (as here) intermixed with third-person address for scientific information. Its tone overall is that of the instructional film rather than the reportage or impressionistic documentary. So, for example, it argues that 'hunger is just as well satisfied by the usual tea of bread and jam as by bread and tomatoes, although it may not do you so much good. So some knowledge of food values is essential for every woman who has the care of a family of children on her hands.' The film proceeds on the unspoken assumption that the root of poor nutrition is ignorance, rather than poverty, with the one exception when Huxley states, towards the end of the film, that 'of course there will – I'm afraid – be many households where the food money won't go far enough. But these rules will help anybody make the best of what there is.' The film combines Huxley's narrative with staged scenes of a middle-aged woman teaching a 'learner cook' how to prepare a casserole, family mealtimes, animated cartoons representing the different food groups – body-builders as a builder, body-protectors as a policeman and body-warmers as a stoker – and a dramatised sequence in which figures personifying different foodstuffs are seen dancing superimposed on a casserole as a means of conveying the virtues of slow cooking.

The reference to a 'plan' in the film's title transfers to the domestic scale the approach that Huxley and his associates were recommending at the national level. The commentary states: 'If you want a balanced diet you'll have to plan your weekly food budget carefully ... You have just seen how all our foodstuffs fall into three main classes and have realised the need for proper planning: planning of shopping and planning of individual meals.'[60] The film closes with a staged conversation between Huxley and Kenneth Lindsay MP, his old associate at PEP, and by this stage Parliamentary Secretary to the Board of Education.

Plan for Living was markedly different from the other two nutrition films discussed here. It is lighter in tone, but still retains the sense of technocracy, of rule by elevated experts, including Huxley and Lindsay. It retains advocacy of the newer knowledge of nutrition – here as the policeman 'body-protectors'. And, in relation to the 'poverty versus ignorance' argument, it is closer to *Enough to Eat?* than the more forceful *March of Time* story, in that it remains ambivalent about the poverty argument, which is relegated to an aside.

By itself, this cycle of nutrition films provides a remarkable example of a prewar fusion of social relations of science and documentary film. Nutrition was certainly the strongest case, but it was not alone. In 1937, the gas industry, working with the documentarist John Taylor, chose air pollution as the topic for one of its two new social concern documentaries. In *The Smoke Menace*, the substantial agreement between particular scientists, filmmakers and sponsoring organisation came together to produce a film that once again laid stress on responsible action by informed and engaged citizens. The representational means used to convey this agreement combined images of industry, domestic life, children, urban landscape and expert voices to produce a whole which could be received favourably as a socially conscious documentary film.[61] Part way through the film J. B. S. Haldane gave a short lecture from a sunray clinic at the Middlesex Hospital on the deleterious effects of smoke on sun exposure – and therefore vitamin D production – of children:

> I am a public enemy. I live in a house with several open coal fires. But I am not sufficiently ashamed when I see the smoke coming out of my own chimney, and I ought to be ashamed of myself. But I know that my chimney is cooperating with a million other chimneys to make London not merely dirtier, but less healthy.

Haldane here was not popularising his own work, but the tone is very similar to that of his newspaper and magazine essays, for example those collected and published as *Possible Worlds and Other Essays* (1927). Here, as in many of the essays, he mobilised an easy expertise, aristocratic authority and self-deprecation as instruments of his scientific rhetoric. For example, he stated: 'Now these children here are getting artificial ultraviolet radiation in order to protect them from rickets. It may be as good as real sunlight; I don't know, I'm not sure, but I do know it's a great deal more expensive.' Within his contribution, it is Haldane's personal responsibility as a citizen, not as a scientist, that is emphasised. The cast of his contribution accords very well not just with citizenship motifs within documentary, but also with Huxley's role in *Enough to Eat?* It is worth considering that in *The Smoke Menace* it was Haldane, ostensibly one of the most radical of scientists, who was found voicing a meliorist, liberal reforming sentiment more typical of the far less radical scientists, though at this stage he certainly considered himself to be a Marxist.[62] This example shows both how intricately we must look at the mediation process in examining vehicles of communication between scientists and the public, but also how complex are the alliances that underpin them. Certainly, as far as documentary films are concerned, nothing more radical than a meliorist reforming viewpoint, normally deployed around ideas of planning,

Fig 3.5 ***The Smoke Menace*****: J. B. S. Haldane in the sunray (ultraviolet light) room at Middlesex Hospital**

was ever expressed regarding any of the public health issues on which they touched. And the emphasis of the science in the public health argument of the film was solidly in the newer knowledge of nutrition vein:

> Growing children need vitamin D, if their bones and teeth are to be properly formed. Now, vitamin D is produced when ultraviolet radiation falls on a waxy substance which is found in almost all living organisms including ourselves. Few children get all the vitamin D that they need in their food, but they can make all that is necessary if enough sunlight falls on their skin. If they don't get enough sunlight or enough vitamin, they will develop rickets and bad teeth. And my smoke is adding to the pall which hides the children of London from the ultraviolet radiation of the sun.

The National Smoke Abatement Society in the five and a half pages concerned with smoke and health in their publication *The Case Against Smoke: The Evidence of Authorities* (1936), did not mention vitamins at all. They concentrated on lung diseases and spoke of the psychological gloom induced by darkness. Yet it is typical of those who championed the use of scientific method in medicine to stress vitamins, which for them proclaimed the validity of the scientific approach. Haldane, who had spent ten years as a member of Gowland Hopkins' Department at Cambridge, was one such advocate (see Clark 1968: 63–4).

Kensal House (1937), made by Frank Sainsbury, was similar in its battery of techniques to *Enough to Eat?* It has a theme of how expertise – in this case modern architecture – could provide the solution to social problems. The Kensal House flats were a GL&CC project, the work of the housing consultant Elizabeth Denby and the architect Maxwell Fry, member of the Modern Architectural Research Group (MARS).[63] Here again was the testimony of both experts and the erstwhile slum-dwellers, the bearers of social problems. The solution in this case was not provided by the state, but by the beneficent action of a corporation, the GL&CC. We may note that, even in the brief catalogue description, M'Gonigle's finding – that the rents on slum-clearance properties may reduce the money available for nutritious food – is subtly alluded to in the phrase 'all this for a much smaller cost than they paid for the inadequate methods of heating and living which they had to put up with before'.

Conclusion

By 1939, through networks of association, documentarists had broadened the range of ways in which science, technology, medicine and the rationality of planning solutions were represented to the public in non-fiction films. The broad range of subject matter selected for documentary treatment by this date featured not just celebrations of the power of science via application in technology but also by virtue of its power to diagnose and cure the problems of society. The filmmakers shared with their associates, exemplified here by Huxley, a commitment to the significance of planning, science (in the shape of the newer knowledge of nutrition) and the transformative role of modern architecture and town planning. The networks of interest that had developed during the 1930s continued into the 1940s. But before they could bear new documentary fruit, they were tested by the state as it adjusted to the enormity of total war.

4 SCIENCE, FILMS AND WAR

> Not only should films about science give facts, but they should present facts in such a way that may well invoke a call to action ... If [they] are going to have real social purpose, they must give people facts and information not drily but dramatically presented. Above all, they must create a desire among the people for science to be used for *all the people*, not for the privileged few alone. (Rotha 1943: 306; emphasis in original)

Different social and cultural contexts produce different kinds of films that represent science, technology and medicine in different ways. In the particular context of World War Two, Paul Rotha took the broader interests of groups concerned with planning and postwar reconstruction and, using his high cinematic literacy, produced a newly forceful version of the socially-conscious science film genre. Effectively, he fused together the genres represented by *Enough to Eat?* and *The Face of Britain.*

The phoney war and beyond: networking consensus on war aims

At the outbreak of war, scientists, planners, architects and documentarists may have shared a sense of self-confidence in the potential social role of applied science and rationality more generally but, as outsiders to government, they had cause to doubt official willingness to take their side. From the time of the Munich crisis of autumn 1938, the *dramatis personae* of this story mutually experienced a lack of involvement in the war effort. They spent much of this time forming groups and associations to discuss how socially useful ends might be served under war conditions. These concerns had a wartime dimension, but from the beginning they were also concerned with planning for the postwar world, with extending the highly organised wartime state into the world beyond war. Each started lobbying from outside government and latterly achieved some involvement in the conduct of war. Along the way, each of these groups envisaged a role for documentary filmmaking. In September 1940, during the Blitz, government began to respond to this pressure by conceding the establishment of advisory committees and by enrolling many of these figures in the war effort. From the beginning of 1941, the pattern for the remainder of the war was set.

Although many of the scientists concerned with the social relations of their discipline had, in the early 1930s, viewed the application of science to warfare as misuse, many – notably the Cambridge Scientists' Anti-War Group – had approved its use to study civil defence, including the preparedness of government for gas attack and bombing. After Munich, many altered their views to promote the use of science for the defeat of Fascism (see McGucken 1984: 155–6; Werskey 1988: 223–34). Both the Association of Scientific Workers and the Royal Society petitioned government for scientists to be granted greater involvement in the war effort. The long sequence of initiatives led by A. V. Hill, biological secretary of the Royal Society, resulted ultimately, and under pressure of the Blitz, in the foundation of a Scientific Advisory Committee to the War Cabinet in September 1940. This was significant in three ways: it raised the official status of scientists in government; brought together representatives of the Royal Society with the heads of the Medical and Agricultural Research Councils and the Department of Scientific Research; and could oversee most science conducted within government (see McGucken 1984: 196).

Architects too experienced a sense of underemployment at first. An *Architects' Journal* book review in August commented that

> many architects have been disappointed at Government inability to make use of more than a fraction of the available skill and experience in the furtherance of the national war effort. A partial justification has, however, always existed in the argument that there is no need for 'architecture' in total war, and that the authorities might be excused for thinking that the profession was mainly concerned in applying embellishments to the solid structure provided by the engineers, forgetting that most of the architect's time is spent in co-ordinating and controlling the work of others, and so could be a vital factor in war organisation. (Winser 1940 134)

Once again, it was under the circumstances of the Blitz that the Cabinet acceded to pressure and appointed a sub-committee on reconstruction (see Young 1981: 89). The appointment of John Reith as Minister of Works and Buildings with a specific brief for postwar reconstruction in October was also welcomed by the planners. It is significant that the department's name changed in early 1942 to include the word 'planning'.

Rotha, in common with many of the documentarists, had a similarly frustrating experience over the first 15 months of the war. At first he was finalising *The Fourth Estate*, his film about *The Times*, after which work dried up.[1] The only new film he made in the whole of 1940 was *Mr Borland Thinks Again*, a five-minute dramatised short for the Ministry of Agriculture via the Ministry of Information (MoI) about silage, released in August (see Thorpe & Pronay 1980: 67). The documentary ideology was predicated on the virtue of the medium as a tool of the democratic state and, accordingly, many film-makers expected their role to be enhanced during the war. But, despite the fact that planning for war propaganda had been underway since 1938, very few opportunities arose for many documentarists during the first 15 months of war. Rotha put it succinctly: 'In the same way as the whole nation became despondent because the threatened blitzkrieg never happened all the winter, so the documentary movement became demoralised because it was not used as the superb piece of propaganda machinery which we all thought it was.'[2] There was a rapid succession of short-lived regimes in the Films Division of the newly formed MoI. Arguments between these officials and the documentarists about what sorts of films should be made became very entrenched. Opportunities for work seemed to improve after a series of appointments in spring 1940, including old documentary allies Jack Beddington (who had commissioned Rotha's *Contact* in 1933) and Sidney Bernstein (manager of the Granada cinema chain, who had financed the documentarists' *World Film News* in the 1930s) (see Swann 1989: 150, 153–60). In May, Rotha wrote and submitted a substantial ten-part *Report on the Production and Distribution of Documentary Films* to the MoI. Amongst the films he proposed 'to maintain public morale and/or to impart information' was a series of six 'about what we are fighting for'. The document is interesting for what it reveals of the development of Rotha's views. He complained about the negative tone of 'win the war today and think about the future tomorrow' propaganda. He emphasised the importance of war aims as he asserted that 'if national morale is to be maintained, it is important to look into the future. It is necessary to give the public a sense of their privileges and responsibilities as citizens of this democratic State and of what they are fighting to protect.' Yet the focus on planning and the role of science and technology seen in *The Face of Britain* is otherwise absent here, except for a proposed second series of six films on the impact of war, which included one on 'how the war has affected Culture and Scientific work'.[3] Much more evident is the typical documentarist's emphasis on citizenship and on 'intimate personal studies' of people caught up in the war. It was this element that persisted with him for the remainder of the year. In the autumn, in the absence of film work, he began to work in an East London soup canteen. He summed-up the situation under the pseudonym 'Documentary' in the *New Statesman* in September: 'By and large, the documentary film movement … has received unfortunate handling by the Films Division [of the MoI]. Its memoranda have been requested and then disregarded, its technicians left for long periods of unemployment, and more than once it was asked to perform impossible jobs by Films Division officials unacquainted with film procedure' (1940: 230).[4] During long months of enforced inactivity, however, Rotha had allied himself to several of the linked groups who were discussing war aims and postwar reconstruction.

It seemed to many of this story's participants that the purpose of the war – beyond the defeat of Hitler – was receiving scant official recognition even up to the time that Churchill and Roosevelt signed the Atlantic Charter between Britain and America on 14 August 1941 (see Calvocoressi *et al.* 1989: 223, 353–4). But, as a result of the period of seeming official lack of interest, the discussion of war aims and of planning for reconstruction became a focus for those whose advocacy of science and planning had been interrupted by the outbreak of war. In some senses, this was a natural continuation of prewar concerns with many of the same personnel, including Julian Huxley, John Orr, Ritchie Calder, Desmond Bernal, Basil Ward and Solly Zuckerman, being involved in several reconstruction groups. Many initiatives and groupings shared characteristics; several, for example, gained an international dimension. The British Association's new Division was avowedly international in scope, drawing on the conflict between the Allies and the Axis powers to promote universalised and internationalised science. Comparable was PEP's change of concerns: 'In the 1930s the focus was on the ills of Britain itself. The effect of the war was electric. It made PEP internationalist but with a single focus, the need for a united Continent led by Britain' (Young 1981: 92).

Political and Economic Planning

The post-Munich period saw Huxley continuing the work of his research committee at PEP, with a study, partially funded by the BA, of personnel and finance devoted to research in Britain (see Anon. 1940d: 5). The war gave added urgency: 'There is clearly a need not only for research into better weapons and better means of protection, but also for research into research. But beyond the war itself still rises the great problem of harnessing science to human needs' (Anon. 1939b: 2).[5] In a short article on 'research and its application', PEP's wartime staff made an explicit link between the need to apply the fruits of research 'in the everyday life of men and women' and the 'scope for fuller use to be made of research results mainly in education or interpretation' via documentary films. Here they cited a joint arrangement between PEP and the Film Centre,[6] funded by the Rockefeller Foundation, to produce scenarios for a series of films showing the impact of the war on social life in Britain; subjects chosen were evacuation, nutrition, food and rationing policy, public opinion and the maintenance of cultural life. The PEP broadsheet commented that 'preliminary exploration suggests that there is room for valuable contact between documentary film producers and social and economic research' (Anon. 1941d: 12–13). Rotha was given the public opinion script to develop (see Marris 1982: 99).[7]

Huxley was active in the Postwar Aims Group, which PEP had established at arm's length from the main organisation in July 1939, and was amongst those responsible for drafting their 40-page discussion of war aims, *European Order and World Order: What are We Fighting For?*, published in October 1939 (see Nicholson 1981b: 47–8). An issue of PEP's *Planning* broadsheet from April 1940 argued that 'Decisions are already being taken in almost every field of social and economic policy that vitally affect the future and dictate in some degree the postwar social and economic structure of Great Britain. The planning of this structure cannot be left to the ... next Armistice Day' (Anon. 1940d: 2). Huxley continued to be involved in reconstruction activities at PEP, discussing, for example, at the height of the Blitz, the re-planning of London with Stephen Tallents and Edward Carter of the Royal Institute of British Architects (RIBA). In his view, rational planning and building were aspects of the social relations of science: 'It is also important to get the scientific side of building represented – the research group of RIBA and the Building Research Centre.' He mentioned that 'Bernal is getting very keen on building research and may go into it wholesale after the war'.[8]

The British Association Division

Most activities of the BA Division for the Social and International Relations of Science fell in the post-Munich period. Its success at sustaining a programme across the war years may well have resulted from its role as a 'popular front' for science, uniting those on the left and centre of scientific politics (see Werskey 1971: 80–1). Those active in the Division, particularly its President Richard Gregory, gradually changed its purpose from the 'objective study of the social relations of science' (McGucken 1984: 124), first to educating the public about specific examples of such relations (for which they chose the medium of conferences) and then to fostering the relationship between science and government as a means to the same end (see McGucken 1984: 147–8). The logic of this approach followed their conclusion in 1938 that citizens in general, and especially their elected representatives, should be considered responsible for the beneficial or destructive use of science; it was the scientists' role to communicate their privileged understanding of their discipline's potential. The subjects of the Division's conferences can be seen as an index of the social relations of science in this period; they also later featured in the documentaries that Rotha and others came to make. In

March 1939 they discussed milk as nutrient and (unpasteurised) as disease vector; in May the mismatch between science and society was their topic; and at the Manchester conference in June Gregory emphasised the importance of promoting clear thinking on science's use, misuse and role in enhancing social welfare. The Association's annual conference in 1939 was curtailed by the outbreak of war. It was not until September 1941 that they held another meeting, a sign of the impact of the uncertainties of 1940. 'Science and World Order' was conceived by Gregory, working with a committee chaired by Huxley and including Ritchie Calder, J. G. Crowther and A. V. Hill (see McGucken 1984: 132–8). Of all their wartime activities this meeting had the greatest impact, intended as it was 'to deal with the relations of science to government and other agencies concerned with constructive planning' (McGucken 1984: 139). Also it was filmed, later to appear in *They Met in London* (1941): 'Spotlights and the whirr of Paul Rotha's camera greeted Sir Richard Gregory as he spoke into the battery of microphones at the opening' (in Gregory 1941: 186). The conference was both outwardly internationalist and, in six sessions, it covered the relationship of science to human needs, world planning, technological advance, postwar relief, government and – in a session chaired by H. G. Wells – 'world mind'. Orr's paper argued that 'a food policy based on human needs would lead to great improvement in human welfare, to a great expansion in agriculture, to increased trade and to the producing of economic stability' (1943: 28).[9] In March 1943 they held 'Science and the Citizen: The Public Understanding of Science', where four sessions discussed 'the exposition of science', 'radio and cinema', 'science as humanity' and 'science and the press', where Rotha spoke. This marked the wartime culmination of their concern with science and the public, though we should note that this was a discussion of ways, means and issues of communication; they spoke amongst themselves, not directly to the public (see Boon 2004b).

The London Scientific Film Association

The Association of Scientific Workers (AScW) – a trades union for scientists with a membership extending to many prominent figures in science, including Huxley – had been instrumental in the Committee on Educational and Cultural Films in 1929. In 1937–38 it established a film committee, with the aim of assisting branches in running scientific film shows. These programmes originally included a variety of Gaumont-British Instructional school films and the full range of documentaries discussed in the two previous chapters, including *Enough to Eat?*, Shell Film Unit works and *The March of Time.* On the basis of the numbers wishing to see the films they established the London Scientific Film Society, with Huxley, A. V. Hill, William Bragg and Lancelot Hogben as its patrons and, at the start, 150 members. It held the first of many weekend afternoon shows, which included *We Live in Two Worlds* and a *Secrets of Nature* film, in November 1938 (see Anon. 1938b, 1938c). The AScW Films Committee set to work reviewing and showing films of several kinds because, as they argued, 'cinematography offers great opportunities for the portrayal of scientific subjects, and a considerable proportion of the output of documentary and educational films covers science from one aspect or another' (Anon. 1939a: 41). Their ideology was clear:

> Science is the foundation of progress. In the long run the benefits of science can only accrue to a society which is capable of understanding and accepting them in their full implications. The basis of an understanding of science lies in the reform of education, but it is almost as important to see that adult minds have the opportunity of appreciating what science is doing and how it is likely to affect human life. Such appreciation can be fostered through the press, wireless and cinema. We hope that the greatest possibilities offered by the cinema in this way will be developed and used to their greatest extent. (Anon. 1939a: 43)

On average, 300–400 people attended their showings in 1938 and 1939 (see Bell 1943: 37). After a reduction of activity in the first months of the war, the committee was active again by June 1940 (see Anon. 1940a: 46) and a year later published a memorandum, *The Scientific Film*, as a key element in a publicity campaign for their work. This argued for the production of both instructional and 'interpretive films', a category which approximates to the dominant concerns of this book, as a film of this kind 'expounds [facts and principles] in such a way as to give them wider significance, especially in their social context. It is essentially a documentary film in which the scientific approach is the controlling motive' (Scientific Film Committee of the Association of Scientific Workers 1941: 3). Films were important for the Association: 'AScW branches can play an important part by running film shows. Apart from their recruiting value, they will be "helping people satisfy the widespread desire to find sense and order in a changing world" and making a step towards the scientific millennium' (Anon. 1941a: 198). Sydney Gregory, reviewing the Science and World Order conference for the *Scientific Worker*, quoting their scientific film memorandum at length, made an explicit connection between planning (linked with science) and publicity:

> It is by practice and example that the majority of the people will be won over to planning and not by airy talk. Moreover, in the long run, it is essen-

tial that the application of science shall be the demand of the majority and not the result of wire-pulling and wrangling on committees. In order that the opinions of the majority shall be favourable to science, scientific workers must take up the weapons of publicity and jostle outworn prejudices and delusions from the minds of the public. (1941: 187)

The AScW continued to promote the use of scientific films and to publish articles about them. For example, in 1943 the scientific documentarist Geoffrey Bell argued that films showing the capacity of science to solve human problems would improve the social status of scientists and thus serve the aims of this trades union. He sought to explain the appeal of scientific films as 'an important aspect of the social relations of science' (1943: 37). The filmmaker 'makes the complex appear lucid – by showing how the radio-valve works, how chromosomes affect our lives, how a bomb explodes – he does something which gives men a great faith in objective, accurate thinking' (ibid.). Linking the continued success of the screenings in London and in many provincial scientific film societies, he announced to readers the AScW's establishment in May 1943 of the English Scientific Film Association, the next stage in the development and promotion of scientific films, which continued well into the postwar period.

The Rights of Man

In February 1940 Rotha discussed with a group founded by H. G. Wells the possibility of making a film on war aims. This dated back to a letter Wells had written at the end of September 1939 as part of a correspondence on war aims in the *Times* that had started the day after the war broke out. Wells had intended his letter to provoke discussion about what the war was being fought for, although it soon merged into his broader concerns with a world state (see Wells *et al.* 1939). The philosopher C. E. M. Joad saw it as being of a piece with Wells' writings dating back to 1915; he teased 'ever since, you have been busy reformulating in different ways this notion of yours of intelligent and public spirited persons, representatives of this mind of humanity *tout court* and generally conceived – God help us! – in the likeness of scientists coming together and salvaging our world' (1940: 155). Precisely what offended Joad appealed to others. Ritchie Calder responded and, in the first months of 1940, became secretary to the committee that drafted the outcome of the initiative, Wells' *New Declaration of 'The Rights of Man'*. Wells and Ritchie Calder were remarkably adept at enrolling influential allies; the committee included Orr, senior physician Lord Horder, journalist and peace activist Norman Angell, Labour MP Margaret Bondfield, Richard

Gregory, economist Barbara Wootton and Francis Williams, editor of the *New Statesman* (see D. C. Smith 1986: 428–49; Ritchie Calder 1941b: biographical note).[10] Ritchie Calder arranged substantial coverage of the debate in the *New Statesman* and the *Daily Herald*, starting in earnest on 5 February 1940. The surviving fragment of his diary from the period reveals the ways in which Rotha and he discussed the possibility of a film on the subject:

> Then came Paul Rhotha [sic] to discuss certain film proposals and told me that Lou Jackson with Oscar Deutch [sic] had had dinner the previous evening with Southwood and Dunbar to discuss a film on the 'Rights of Man', which Lou Jackson thought the greatest title since 'Birth of a Nation'. Paul says that whatever should be done, we should retain the drafting committee, as in fact a scenario committee, so that nothing would be done which would infringe the spirit of the debate.[11]

The rights of man enterprise spawned a series of pamphlets – including three published by Penguin – and nourished the concerns of Wells' twilight years. It also cemented relationships in Rotha's and Ritchie Calder's circle.

The 1940 Council

A significant group for Rotha, concerned with the built environment, was the 1940 Council, an alliance of architects, planners and scientists interested in war aims. It was founded at a meeting under the title 'Problems of Social Environment and the War' convened by the Housing Centre at RIBA in February 1940. The chief objective of the meeting was the formation of an autonomous council of 'not less than 30 members' to watch planning, architecture, housing, economics and social services 'nationally, co-ordinating and assisting the work of existing bodies' (Anon. 1940b: 132). Edward 'Bobby' Carter, librarian at RIBA, was a key member of the group chaired by Lord Balfour of Burleigh which included Conservative politician Sir Montague Barlow.[12] The membership was highly influential; Burleigh was leader of the parliamentary group promoting town planning and Barlow had been Chairman of the Royal Commission on the Geographical Distribution of the Industrial Population, which reported in January 1940 stressing the need for national planning to redress the problems of the unbalanced industrial population (see Hasegawa 1999: 139). The Council's immediate objective was 'to promote, through research groups and by other means, the planning of social environment on a national scale and to make widely known the need for such planning' (Anon. 1940c: 156). Carter maintained that the Council was a broad church appealing to 'people from Balfour to Bernal who respond to the same pattern of ideas'.[13] As with the BA and PEP, 'giving the people the facts', as Carter said, became its focus because, he argued, 'schemes for planning must grow out of the people's awareness of the need for improvement of their conditions ... until the people are aware, they cannot delegate their authority to the experts to work things out for them'.[14] This group welcomed John Reith's appointment as head of the Ministry of Works in October 1940, with a brief to address issues of reconstruction and planning (see Stevenson 1986: 69–70). Reith appointed a Committee under Mr Justice Uthwatt, which in 1942 successfully recommended the establishment of a central planning authority and rights for the state to compulsorily purchase land for development. Reith also established a consultative panel of experts, including several from the 1940 Committee, on physical reconstruction in February 1941, and this had considerable influence until Churchill dismissed Reith a year later (see Hasegawa 1999: 142–7). But the attitude of civil servants shows the extent to which planners, as much as scientists, were involved in 'outsider politics'. One wrote that panel members should be persuaded 'how they can be of service to us, namely by being at our service when we want information and advice and leaving us in peace when we don't' (quoted in Hasegawa 1999: 142).

Rotha first had contact with this group early in October 1940. He immediately saw the potential of making a 1940 Council film for a return to documentary's purpose of 'using film as a form of social analysis'.[15] He met Carter a few days later and wrote in his diary:

> A long and stimulating talk with E. J. Carter, librarian at RIBA who seems to be mixed up in a lot of interesting things. Explained setting up of the 1940 Council. Nobody on it with knowledge of public relations. He is to try for the setting up of a Sub-Committee with full powers to co-opt outside + explore the film position. The Council ... will have two aims: research material now for future planning: publicity to arouse consciousness among ordinary people of their needs for proper social environment. Films to be used in latter campaign.[16]

We should note Rotha's language here. He was concerned with the needs of ordinary people for 'proper social environment', that is, one planned by appropriate experts, but there is no explicit reference to scientists. But Carter and the 1940 Council became key contacts for Rotha in the years that followed, and science became a significant element of his concerns.

The Tots and Quots

Moves to re-establish the Tots and Quots group had commenced at the time of the Munich Crisis 'to alert scientists to the dangers that lay ahead' (Zuckerman 1978: 396). The first reconvened meeting, in November 1939, discussed the way that the scientific establishment, not wanting to disturb the politicians and civil servants running the war, were disparaging those scientists who wanted a hand in the war effort. Regular attendees included Bernal, Ritchie Calder, J. B. S. Haldane, C. D. Darlington, Richard Crossman, Edward Carter, Roy Harrod and Solly Zuckerman. The role of scientists in the war and postwar reconstruction were the dominant themes of the meetings (see Zuckerman 1978: 109–10, 396–404). Sidney Bernstein from the MoI Films Division was a guest at a meeting in September 1940 when Tom Harrison commenced the discussion with some observations on the public image of scientists (see Zuckerman 1978: 400). The group's most visible product was the publication in July 1940 of the Penguin Special *Science in War*, written by 25 scientists and allies.[17] Allen Lane, present at the dinner of 12 June 1940, had 'accepted the challenge' to publish a text based on the discussion that evening, if it could be produced within a fortnight. Zuckerman's group accepted and the book came out in July. It argued that 'the issue of the war, and the nature of the ensuing peace, will largely depend on how effectively and how quickly science can be used' (Zuckerman 1940: 142). In a discussion of propaganda on behalf of science, one of the contributors made reference to scientific film societies and complained that

> The scientific film has not yet caught the public interest except in the dramatised form typified in the film about Pasteur [a reference to the 1936 MGM biopic]. If more money could be spent on the production of films intermediate between a formal documentary and a romantic life history, it should be possible to tell the story of the past achievements and applications of science, and to show something of the activities of the research workers today. (Zuckerman 1940: 125)

This comment echoed Huxley's criticism of *Enough to Eat?*, seeking to use dramatic technique to render scientific films more compelling to the viewer.

Just a few days after his first meeting with Rotha, Carter arranged for him to see Zuckerman and Darlington, whom Rotha knew to be amongst the authors of *Science in War*. Rotha enthused in his diary:

> Got down to discussing films as a medium for publicising scientific progress ... Z very alive to possibilities; Carter anxious to formulate plans. As

> we talked, it became clear to me that there is an immensely important opportunity here of getting some of these specialists, each brilliant in his own field, into a group and setting them about the job of planning for the World Beyond War.[18]

Although he had known several members of the Tots and Quots since the mid-1930s, this meeting with Zuckerman was the true beginning of Rotha's engagement with scientific and medical documentary as a specific category that reached fruition over the next several years. He continued:

> If we take all the fields most closely affecting social environment – nutrition, medicine, agriculture, economics, social science, architecture, etc – + picked out the most progressive worker in each, added some of those who are working in touch with 'public opinion' we could form a group of people who could formulate a plan for better living. That plan could be publicised - by press, film and radio, with the intention of creating the need for its being carried out when the war is suspended.

He spelled out the role of publicity – by which he primarily meant documentary – in the revolutionary implications of such a scheme under the special circumstances of war:

> If we can create an awareness in the minds of people of how much better they could live their lives if the full resources of science, industry, agriculture + architecture, etc [were made] available and made use of, nothing would stop those people getting them. But the resources have got to be known, not only the resources but their applications + publicity is necessary for this vast job. It can be done – + now with the stimulus of war to urge the necessity of a progressive peace, it must be done.

This will to create public awareness of the world that science could make became highly significant for the subjects Rotha chose to represent and, once he had the opportunity, he focused on producing films that conveyed these themes.

Rotha films: social aspects of the war and planning for the postwar

The changes in staff at the Ministry of Information culminated for the documentarists with the appointment of one of their number, Arthur Elton, as Film Officer in January 1941 (see Sussex 1975: 121; Swann 1989: 150–60). At the same historical moment, Rotha set up a company and started making films again.[19] He registered two company names, the plain 'Paul Rotha Productions' and the punchy 'Films of Fact'. The word 'fact' in documentary discourse relates to claims of representing reality. However, in Rotha's case we may perceive an extra dimension because by this date he had developed a new level of engagement with planning and science, which for him expressed rationality applied to address the social issues that were the conventional concern of documentarists. His company's first projects included MoI commissions on salvage, blood transfusion, school education and diphtheria immunisation. The possibilities opened up by his new associates were uppermost in his mind as he recorded in his diary in January that 'the MoI films – those that we choose to accept – will be a sheet-anchor – but I want to explore the postwar field of reconstruction. There is the Science in War group plan + the "1940" Committee to pursue: and this interests me a great deal. Here is a real social growing point for documentary.'[20] The press release announcing the company stressed that 'documentary film is a medium through which audiences both at home and overseas may be shown how the Britain-to-come may be discussed and planned even while a war is being fought'. Rotha continued:

> The Unit will establish a policy of producing films about progress in the fields of education, health, medicine, housing and the social sciences in terms of both war and postwar problems. Its research will be directed to this end, being based on work conducted over ten years in co-operation with specialists in these fields.[21]

He published an advertisement in *Documentary Newsletter*, which carried a quotation from Julian Huxley's *The Uniqueness of Man* (1941): 'The most vital task of the present age is to formulate a social basis for civilisation, to dethrone economic ideals and replace them by human ones' (Huxley 1941: ix). The advertisement continues:

> We subscribe to this statement; and we believe that the medium of the film should be used to spread the knowledge needed by all who plan for the postwar world ... We believe that today it is the responsibility of documentary filmmakers to produce films and conduct research about those aspects of our life from which will come social changes in the future. (Anon. 1941c: 59)

Rotha had propagandised on behalf of scientific planning in the prewar period, but given its ubiquity in the 1930s, not excessively. But his strengthened contacts with those promoting planning and science in 1940 had the effect of intensifying his convictions in these areas, building them into his sense of the social role of documentary.[22]

The contacts continued; in early February he recorded that he had 'Lunched with Carter from RIBA. Maxwell Fry, architect, + Tom Wintringham, militarist [the *Daily Mirror*'s military correspondent], in attendance. We discussed the possibility of an advisory committee to Rotha Unit – of experts in the fields in which we specialise. People like Zuckerman, Huxley, Haldane and others.'[23] At the end of the month, he attended a Tots and Quots meeting, where, to a gathering including Zuckerman, Bernal, Carter and Huxley, he voiced his sense of the need for the MoI to move on from films 'of the "carry on" variety' towards those concerned with postwar reconstruction, and the need for an advisory body to the MoI to achieve this.[24] 'The obvious use of the documentary film for interpreting scientific work was explained and, I feel, useful contacts were made.'[25] In early March, Rotha arranged for the 1940 Council to form a sub-committee to deal with films about reconstruction.[26] One of Rotha's first actions was to show this group *The Face of Britain*, together with *The City* (1939), an American film directed by Ralph Steiner and Willard Van Dyke, which may have derived its combination of dialectical structure and treatment of themes from Rotha's film.[27] He wrote to Grierson:

> There are others whom we know like Calder who are doing a good job. Huxley is forever restless flitting from one place to another. Quite a few of the scientificos are making a very good job of things at the moment. PEP is much operative. I am keeping close company with these people these days. They have much to give in the way of specialist knowledge. That is why I have formed the committee about which I told you.[28]

Rotha's contacts with Ritchie Calder continued throughout this period (they were both living at Clifford's Inn). A diary entry records 'broken off here because Ritchie Calder comes along with his two new pamphlets "The Lesson of London" + "Start Planning Britain Now"'.[29] The latter was concerned to show that the destruction wrought by bombing could provide the opportunity for improved living conditions, and he made the broader connection: 'physical planning must not be separated from economic and social planning or from the cultural and the ethical' (Ritchie Calder 1941a: 20). He saw the London of the Blitz as the launch pad of revolutionary social change. *The Lesson of London* is more vividly rhetorical:

> We are not fighting for 'Democracy' … with a capital 'D' … not for the 'Democracy' which was once a coin brightly minted from the gold of men's

> aspiration, and to which the French Revolution gave its superscription 'Liberty, Equality and Fraternity', but which has been so long in the pockets of the Ruling Class that its device has worn off ... not for a Democracy of privileges and slums. No, the people who are holding the Front Line are fighting and suffering for a new democracy, which they can understand, and to which they are giving the meaning – 'democracy' without the capital. The coin is being reminted, its true worth fresh assayed in the fires of human endurance, and its superscription is that simple remark of the docker: 'We are all in it together.' (1941c: 124)

Rotha's left politics expressed themselves in a particular interpretation of reconstruction that included a strong sense of social revolution, as he privately noted in his diary: 'The big industrialists and manufacturers will only go so far in the war effort so as to secure a so-called "victory" – but they still care [most] for profits ... The money makers will ... prefer a negotiated peace, in which they hope to retain their money and their status, rather than a peace in which a more equalised method of living will be demanded.'[30]

Science and War: the film that never was

Rotha's contacts had their most explicit result in a film Rotha planned with the title *Science and War*, first mentioned in his diary in May 1941, for which first draft scripts were being composed in June 1941.[31] This film provided the template for several of his later major works, both in terms of the structure and its subjects. He sketched the socially-useful complexion he put on science:

> There's a biggish [film] on Science and War and just how science has been given its chance in Britain since September, 1939. And by science I don't mean guys splitting atoms or even getting a fighter to fly at 400 instead of 399 mph. I mean the fact that the population is for the first time in history getting an adequate diet, that agriculture is being approached scientifically for the first time, that free milk is being given to millions of kids, that inspection of bomb damage is revealing facts about building construction that will vitally affect reconstruction, that from now on science must be an integral part of society ... The technologist, the scientist, the architect, the specialist has at last been called in by the British state. Those are the men who are going to plan for a new Britain ... if this film can educate the great mass of people just what resources of science lay ready for the common good, perhaps the great mass will see that they are made use of.[32]

He intended that the film should end with sequences of the Science and

World Order Conference. Serious work was done on the film; for example in July Rotha visited the Air Raid Precautions branch of the Home Office where old associates Basil Ward, J. D. Bernal, James Baker and others were working on the design of air raid shelters.[33] In the same diary entry he commented that *Science and War* 'grows more difficult every day. It's not one film but fifty. But the secret is not to be led away by the potentialities but *to make the film that will have as its aim the relating of science to the ordinary person*' (emphasis added). He secured, via Basil Ward, permission to film experimental tests on shelters and dreamed of a government secret scientific film unit.[34] Huxley, Bernal and Hogben assisted him with the general theme of the film[35] and, by early August, W. G. Bennett and Michael Orrom, working for his company, had completed a first treatment. Rotha sent it out for discussion to a wide group of people, including Carter, Crowther and Zuckerman.[36] The treatment lays out the film's theme:

> Science needs freedom; but freedom needs science. The use of science for social progress is controlled by material circumstances. Freedom of research, of experiment and application, depends on Government and Industry's attitude towards the community and, reciprocally, the attitude of the community towards science. In war, science is aiding both protagonists, in offence and defence. From out of war may come; either a misuse of science to control the community for anti-social ends, or international cooperation in the scientific method to achieve the social well-being of all peoples.[37]

The film was to have had a three-part historical dialectical structure, which was intended to start with a prologue, an introduction to the film locating it in the contemporary state of affairs of being at war. But the treatment represents this as a war of ideas (liberty and reason against repression and irrationality); of productive power in offence and defence; and only finally of killing. This initial section was to have ended with a summary of the offensive and defensive use of science by the allies against the Nazis. In the first main section – the thesis of the film – Rotha compared the prewar state of science: in Britain, where, echoing Huxley, the organisation of science was described as 'haphazard'; in the USSR, described as having 'the first fully planned system of State scientific research'; and in Nazi Germany in 'the service of the state and its ideology'. In Britain, it contends, academic science had been productive and popularisers had been active but – as in the case of James Jeans (the astronomer and populariser) – with a tendency to mystify. Quoting Crowther it criticised the majority of scientists for not accepting science's social relations. But, the treatment argued, science must be applied in three ways: in technologies ('the science of the industrial laboratory making lifeless things serve man'); in natural processes (the example here is agricultural research) and 'to man himself' (where the instances include Orr's *Food, Health and Income*, plastic surgery and the antibacterial drug M&B 693). Moving to its second example, the treatment says that the Russian state has stimulated research and application; it gives, amongst its examples, Lysenko's vernalisation,[38] synthetic rubber, blood transfusion and rural electrification. But it warns that 'State control can also retard qualitative scientific progress on ideological grounds (controversy on genetics)'. In Germany, where science is subservient to war, 'the Nazis deny science because it involves free thought' and 'a biologically false race theory is put forward'; 'The Nazis maintain irrationality but at the same time make full use of science in the technology of war' and are 'fully alive to the importance of nutrition'.

The film's antithesis was to look at science in war in Britain, starting with the rhetorical elision that 'from the outset of the Second World War, the British Government has realised the prime importance of scientists in the war effort'. The treatment explained how some scientists were recruited to research new offensive and defensive weapons, others to produce them on an industrial scale or to be trained in the necessary skills with results that, though useful in war, will also have peacetime value (plastics is one example). But, 'of far greater long-term importance is the calling in by Government of scientists to look after the physical well-being of the population'. Citing the nutritionists John Orr and Jack Drummond, it quoted Huxley: 'necessity coupled with propaganda campaigns are teaching people to adopt a diet which is both cheaper, better balanced and richer in vitamins'. Communal feeding, free milk and school meals for children were stressed, as were improvements in agriculture, health centres, blood transfusion, burns treatment, thoracic and plastic surgery and nerve grafting. But, the film stressed, 'the war has merely hastened large-scale experiments upon ideas and material which were established before the war'.

The film's synthesis looked towards the future:

> It is becoming increasingly obvious that the integration of scientists, the people and Government will continue after the war [which] has shown already that the time lag between research and application can be appreciably shortened. The war is already showing that when a new Britain is planned, there will be a wealth of scientific knowledge available for reconstruction. Reconstruction must inevitably be planned by all branches of specialists and trained men, including all branches of scientific workers, who are today gathering data for their plans. The spirit of science will underlie this planning for society.

This section listed various reconstruction groups – including the 1940 Council – that 'between them embrace the fields of food policy, population trends, medical service, conservation of natural resources, housing and agriculture'. A reference is made to the Uthwatt Committee's 'vague' commitment to central planning. A section deals with increasing international co-operation in science: in person, in data and in publications, a type of collaboration symbolised by the Science and World Order conference. The outline's ringing final phrases read: 'Co-operation between scientists and the governments of peoples and between scientists of all nations appears inevitable. As the ideological war loses foundation, as nations become more co-operative, internally and externally, the way should be clear to a scientifically planned approach to social wellbeing.'

Science and War never went beyond this outline, for reasons that are difficult conclusively to establish; most likely it was simply one of the many films dropped at first treatment stage.[39] Had it been made, it would certainly have been the most focused fusion of social relations of science discourse and documentary film. However, its concerns lived on in later projects. In mid-August, Rotha suggested to the MoI that 'they allow me to make a short [film of the Conference] for immediate priority release as well as sequence for "Science and War" film. If so, we will use Ritchie Calder as narrator and compere to explain Meeting.'[40] In the end, *They Met in London*, the 12-minute film released early in 1942, was a documentarists' version of a newsreel, distributed by the newsreel company British Paramount News, voiced by John Stagg. It hints at the scope and ambition of *Science and War*, offering the first public cinematic statement not only of Rotha's and Ritchie Calder's views on the role of science in war and beyond, but also of the several groups described above.

Rotha and Neurath

In parallel with the development of *Science and War*, Rotha began working with two figures who reinforced his commitment to planning and to the significance of the social relations of science, and who became practical collaborators – Otto and Marie Neurath, newly in Britain after fleeing the Nazis in Austria and Holland. Otto Neurath had long-standing experience of social and economic planning from his involvement in the Bavarian revolution and in interwar 'red Vienna' (see Neurath & Cohen 1973: 18–28). Neurath's book, *Modern Man in the Making*, argued that planning was becoming universal, and stressed its deep historical roots; 'only its widening scope and our being conscious of it are new' (1939: 86). Anticipating arguments about freedom from want, it also argued that 'if a planned economy were a universal institution, the making of a good livelihood might then satisfy all important human needs' (1939: 87). His connections with the architects of the Bauhaus speak of his concern with the rationalisation of the built and designed environment. The more the Bauhaus had emphasised technical and scientific approaches and espoused a social drive for their work, the more Neurath believed them to be espousing with him the 'new form of life' (ibid.), effectively an expression of the Modernist cast of mind discussed in chapter two (see Galison 1990: 715–16). In his Isotype system of statistical representation – a type of universal picture language – he had developed a visual means to promote social and scientific planning to the general public. (Isotype is an acronym of 'international system of typographic picture education'.) This was a system that entailed following a set of rules in the manipulation of a set of standardised symbols. The core principle was that increased quantities were always represented by multiplying the numbers of these symbols, not by altering their size (see Reisch 1994: 153).[41]

The Neuraths, released from internment in February 1941, had settled in Oxford under the protection of G. D. H. Cole, the Fabian economist, with the intention of restarting their Isotype work. Alerted to their presence in Britain by Ritchie Calder, Rotha wrote to Otto Neurath regarding possible work on the blood transfusion film: 'Having admired your work in pictorial symbolism for many years, I feel that there is no-one more qualified to help in this matter than yourself.'[42] Neurath responded enthusiastically about the use of Isotypes in films, with which he said he had previously experimented.[43] Rotha made a visit to Oxford in May 1941.[44] The auguries were good for the meeting, given Rotha's strengthened interest in the social role of science and planning and Neurath's established commitment to the same principles.

Rotha and the Isotype Institute collaborated on 17 films during and after the war (listed in Edwards & Twyman 1975: 45). Some of these, including the five-minute salvage film *A Few Ounces a Day* (1942), were pure exercises in the use of Isotypes; others were on explicitly scientific themes, such as *Blood Transfusion* (1941). Just a few expressed the themes intended for *Science and War*. One of these was a film that was extensively discussed under the title *I am a Scientist* (1942), which sought to present a human face for science whilst conveying its social usefulness in a way that would have resonated with Geoffrey Bell and others active in the London Film Society.[45] Neurath, with his long-term interests in visual communication, became particularly interested in developing types of films that blended Isotype and documentary approaches; this led to the high level of engagement that is evident in the part played by these animated diagrams in the films they made together.[46]

World of Plenty

The documentary that brought together many of Rotha's concerns and contacts was *World of Plenty*, made with Eric Knight and John Orr between autumn 1941 and spring 1943. This film, which will receive extended treatment here, was significant in several ways. It employed newly forceful cinematic technique to create a new genre for the representation of science and public health that fused social concern about adequate feeding with scientific planning solutions. For Orr, and others who had been involved in discussions about the social relations of science, it marked a new stage in their public rhetoric. Where *Enough to Eat?* had been concerned solely with the British case, this new film took the world as its subject, in a way that was congruent with the wartime cast of interests at the BA. Confident in the favour shown to planning during the war, it imagined a postwar world where the same techniques could be applied to solve the global problem of food shortages. In cinematic terms it turned into Rotha's most sophisticated, large-scale and consistent film up to this date. It used dialectical montage, but in a film of much more significant scale than before; at 43 minutes it is more than twice the length of *The Face of Britain*, for example. This was combined with multi-voice commentary, here used much more extensively than in *New Worlds for Old* as in this case it was used as the key structuring device of the film. Whereas in that earlier example the sceptical and questioning voices simply introduce new stretches of narrative exposition, here the cut and thrust between the protagonists acts as a dialectic in sound to which the image track is subordinate.

The commentary is primarily an argument between two central protagonists who, like all the main voices, are not seen on the screen. The first voice, 'Newsreel', spoken by Gaumont British News commentator E. V. H. Emmett, was described in a draft script: 'We've heard him before, on lecture platforms, at election meetings, on the newsreels; the voice of authority, fluent, unhesitant, but so often wrong – before, during and after the event.' The other main voice, 'Man-in-the-Street', spoken by Eric Knight was 'puzzled, critical, sceptical but eager to know about the chances of a fuller life; perhaps a little slow on the uptake'.[47] An example of their interplay occurs near the beginning, where a series of animated diagrams is used; the script reads:

> Newsreel (slowly): On the left are Britain's imports per day in shiploads. On the right you see what she produced for herself. The two added up to her daily consumption of food ... National food needs thus formed one of the basic elements of a strong flow of healthy world trade.

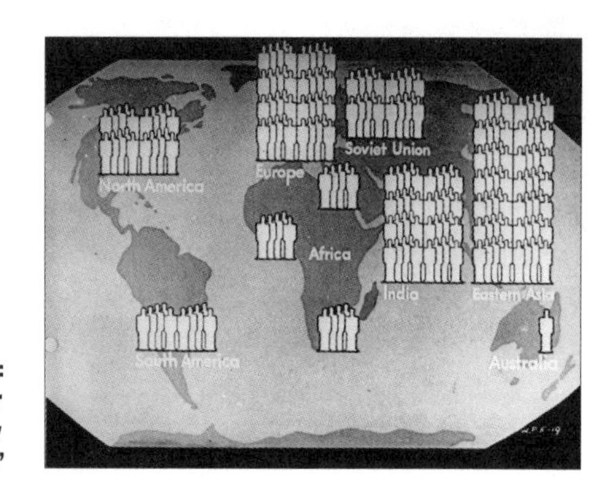

Fig 4.1 *World of Plenty*: Isotype – 'Give me half an hour with Walt Disney and I could pay my income tax and never feel it'

> Man-in-the-Street (cynically): Healthy world trade! Yeah, you can prove anything with diagrams. Give me half an hour with Walt Disney and I could pay my Income Tax and never feel it. (Knight & Rotha 1945: 10)

This playfulness, relishing the comic potential of the technique whilst remaining earnest about the film's argument, is one of its hallmarks. In addition to these key protagonists, there are two commentators (voiced by Henry Hallatt and Robert St. John) and a handful of smaller parts for unseen announcers. Several other roles are acted, including British and American farmers and a British housewife, played by Marjorie Rhodes. Several experts are seen, most prominent of whom is Orr. New shooting was undertaken for the animated Isotype diagrams, but around two-thirds of the visual content was footage carefully selected from over a hundred other films.[48]

World of Plenty tells the story of food production and distribution before, during and after the war, using the structure Rotha had intended for *Science and War* of a three-part historical dialectic preceded by an introduction. It opens with a magisterial prologue, over shots of grain fields: 'This is a film about Food. The World Strategy of Food. How it is grown – how it is harvested – how it is marketed – how it is eaten. In peace or war, Food is Man's Security Number One'. The film's thesis, 'Part One: Food – As It Was', concentrates on the prewar food production and distribution in Britain and America. The state of the populations of these countries during the Depression is discussed, with the audience being carried by Man-in-the-Street to focus on the injustice of the situation; he says America is 'the wealthiest nation in the world and a third of it ill-nourished! Laugh that one off with a diagram, mister!' An American farmer is introduced to explain how artificially high market prices produced a glut of farm produce: Newsreel suggests a programme to restrict the amount of food grown, but is interrupted by a sequence contrasting food destruction and human needs (quoted below). This is followed by an explanation of income distributions in America. There is a comparison with Britain, labouring under similar agricultural policies, partly voiced by L. F. Easterbrook (whom Rotha had recruited to be on the reconstruction committee of the 1940 Council). This section features a British counterpart to the American farmer. John Orr interrupts the discussion about whether or not British people looked undernourished, with a sequence on the newer knowledge of nutrition, British income groups and reductions in the incidence of rickets and tuberculosis. Orr states that at the end of the 1930s, 'We were just on the brink of making great progress. We were going to have a world…'. He is interrupted by the sound of shooting which brings the first part to a conclusion.

The film's antithesis, 'Part Two: Food – As It Is', starts with Newsreel, Second Announcer and Man-in-the-Street presenting an exposition of increased British food production and the problems of shipping food in wartime. An excerpt from President Roosevelt's speech of 17 March 1941 introduces Lend-Lease, under which the American Government purchased goods required by Britain from American suppliers in assistance to the Allied war effort. A dramatised discussion between Man-in-the-Street (off-camera) and a 'typical British housewife' explains rationing policy. Man-in-the-Street, concerned about the size of the rations, questions Lord Woolton, Minister of Food, who provides further details of food in wartime. An 'Englishwoman' explains how expectant mothers and children are given special supplies. Man-in-the-Street wonders 'if all this talk about vitamins is just a fad'; Lord Horder, a senior medical figure, explains that the health of Britain is 'remarkably good: better in fact than we expected'. The Englishwoman explains the potential and actual benefits from improved feeding of children and mothers.

'Part Three: Food – As It Might Be', the film's synthesis, starts with an animated diagram clarifying the previous sequences:

> Second Announcer: We have seen: the State directs food to the Mines and the Factories so that the workers are fit and strong to make the weapons of war … and to the Fighting Forces so they are fit and strong to use the weapons of war … To the Youth of the nation to ensure vigorous man-power … To the Children to ensure vigorous youth … To the Babies to ensure healthy children … to the Mothers and Mothers-to-be, to ensure healthy babies … the circle is complete and the relation between the Individual and the State is clear. An individual's duty to the State is to keep healthy. The State's duty to the individu-

Fig 4.2 ***World of Plenty*****: proposing citizenly rights and responsibilities in relation to food supply**

al is to ensure the *means* to keep healthy.

Newsreel: Except that in the rush hour of war, we've had to work backwards, from adult to baby. And in the peace to come it will work forward – healthy mothers – healthier babies – healthier babies – sturdier children – sturdier children – more vigorous youth – more vigorous youth – stronger adults. Stronger and healthier and happier!

This sequence is highly significant for the film's argument because it not only equates health with good nutrition, but expresses the problem of malnutrition in the documentarists' familiar language of citizenship, as subject to a type of contract between state and citizen, a contract which is presented as the basis of postwar policy. After this, Man-in-the-Street introduces a sequence about world food; another American farmer and his family listen to Claude Wickard, Secretary of the US Department of Agriculture, talking on the radio about the need for increased production for the Allies. Then comes an exchange featuring the American and British farmers from section one. They demand to know whether, in the postwar world, they will be 'left holding the bag' like their fathers after the previous war. A further quotation from Wickard answers the American farmer, and the British farmer is answered by an excerpt of Orr speaking at the Science and World Order Conference: 'In this country war has forced us to adopt a food policy based on nutritional needs – we no longer produce what the farmer thinks will give him the biggest profits, we produce the food we need to feed the nation'. Orr mentions the Atlantic Charter, which set out the war aims to be pursued if America were to enter the war; the film picks up and extends its sixth point: 'Freedom from want means food for everybody on the new gold standard of health.' Sir John Russell, Director of the Rothamsted Experimental Station, talks of the need to restore food production in enemy-occupied lands. First Announcer states that the Nazis in retreat will 'strip the land bare'; he goes on to state explicitly the scientistic assumption of the film:

> 'Science has the answer! We know now what can be done! Munition factories must be changed over to make farm machinery. Experts can say what kinds of seeds should be sown; what kinds of fertilizers should be used. And today we have artificial aids for the breeding of animals. Here Russia has already shown the way. A bull in England can be mated with a cow in Poland without leaving its home field. In this way the best strains of

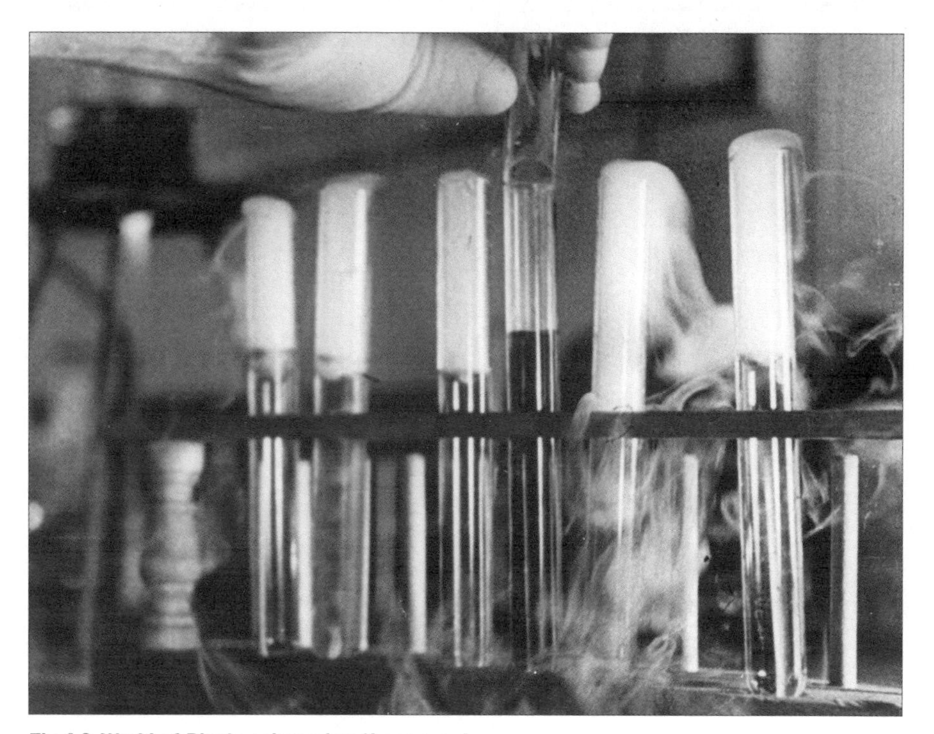

Fig 4.3 *World of Plenty*: science has the answer!

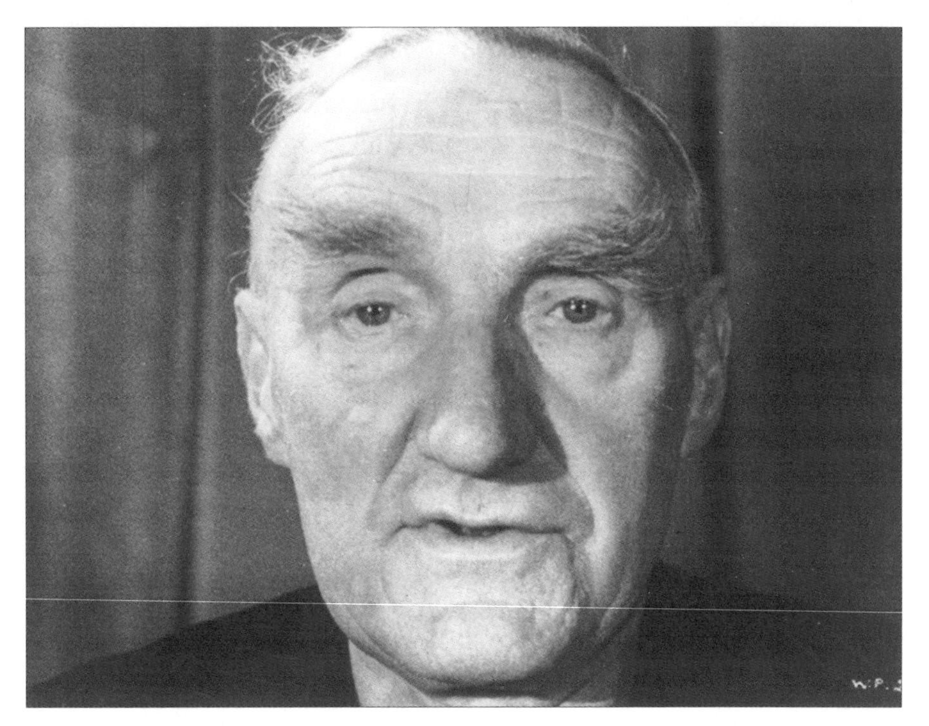

Fig 4.4 ***World of Plenty*****: John Orr: 'What are we fighting for if not for something revolutionary?'**

> livestock can be sent to the peasants of the poorest countries. Our scientists already have the knowledge by which the devastated farmlands can be made productive again.

Orr reiterates the theme that prewar food production had not been geared to the health needs of populations, and First Announcer hammers home the message: 'Iinternational trade doesn't make sense unless it is based on supplying human needs by making the resources of the whole world available to all the peoples of the world.' Second Announcer takes up the refrain with talk of the necessity of a world food plan. Newsreel bursts out after this sequence, 'But this is revolutionary!', to which Orr replies 'grimly with great feeling':

> Tell me, what are we fighting for if not for something revolutionary? What do people like you and me hope to get out of this war if not a better world? The empty slogans 'A World Safe for Democracy', 'A Land Fit for Heroes' – These mean nothing! Plain people know what they want. They want security … The Common Man everywhere demands freedom from want … We cannot attain freedom from want until every man, woman and child …

shall have enough of the right kind of food to enable them to develop their full and inherited capacity for health and well being.

The film ends with an excerpt from Vice-President Henry Wallace's 'Century of the Common Man' speech, concluding with the statement: 'There can be no privileged people.'

The language of freedom from want became a powerful rhetorical and organising principle for Rotha and his associates. In his book *Fighting for What?* (1942) Orr advanced in popular form an argument based on this principle of freedom from want (Rotha's 'what human needs require'), an explicit citation of Roosevelt's 'four freedoms' speech delivered to congress in January 1941, which was a significant stage in America's increasing involvement in the war. (The other three components related to freedom of speech, expression and worship, and from fear.) Orr argued that 'men of science have now given us the physical means of attaining the new world order which our poets and prophets have seen only in their visions' (1942: 11). This sense of direction in society required a familiar type of political action:

> We must draw up a definite plan indicating the steps by which we are going to reach the better world we aim to build ... We know what man needs, for his physical well-being, food and shelter – a house, furniture, clothing and a job. And the standard of food and the standard of housing necessary for health have been defined. (1942: 9)

It would be entirely correct to perceive here a resonance with Huxley's developing line on the need for scientific research to be attuned to the needs of consumers, which fuelled much of *his* interest in physical planning and reconstruction, which he was exploring via his involvement with PEP.

World of Plenty derives its impact from several aspects of its technique: the multi-voice commentary, William Alwyn's music, and from the care with which sequences were shot, selected and edited together. For example, the convention of movement to the left signifying America, and movement to the right signifying Britain, is rigidly and sometimes playfully sustained (see Pearson 1982: 75). It is difficult to convey in words the film's passionate tone, although contemporary viewers certainly praised its effectiveness. The *Times* review, for example, stated:

> The audience are left with the impression that they have been treated to a frank, realistic, and pictorially lucid discussion of a complicated problem. It is important to note, however, that the film would not have served this or any other useful purpose if it had not contrived, whilst treating a serious

> subject seriously, to keep expectant and amused the spectator's pleasure-loving eye. (Anon. 1943c: 5)

World of Plenty, like all the films we are discussing, was a particular contingent product both of the interests of those closely involved in its production, and of its circumstances, especially the interests of the several ministries on whose policies it touched. A decade after production started on the film, Rotha maintained that

> Of one thing I am convinced, only four of us – Orr, Calder-Marshall, Knight and I – understood the full scope and scale of the film as it eventually turned out. Officialdom saw it merely as a dull, dreary, non-theatrical picture about food rationing. (Quoted in Knight 1952: 198–9)

Contemporary sources, however, show how broad the network of people involved in the film's production was, shaping its form both by supporting Rotha or by opposing aspects of the film's script. Rotha's contact with Neurath in this period was certainly significant; his diary and letters show how deeply impressed he was:

> The great thing about Otto Neurath is that he is a brilliant economist first and a diagram king second. The more I talk with him, the more I am impressed with the man's brain. His ideas on planning are sane and inspiring. He has been at all of this for twenty years and, in his own field, is as brilliant as Orr and the others.[49]

But, however great the fellow feeling, Neurath's specific impact on the film's argument is difficult to discern, beyond his supply of a significant number of commissioned Isotypes, although Rotha maintained that he was 'a contributor of ideas to the film throughout'.[50] The core individuals brought particular concerns, but it is also demonstrable that they had a significant range of shared beliefs: in the pressing nature of the social problems of the interwar years; the relevance of applying scientific and planning solutions to address social needs, especially in nutrition; the importance of developing an active and engaged citizenry; and the role of publicity and especially films to ensure that end.

The first move in the film's production was a letter from Rotha to the Ministry of Information proposing a film on the strategy for food in October 1941, very soon after *Science in War* must have been cancelled. Citing the 'important part food resources and conservation played in the BA meeting and how food is tied up in the lease-and-lend agreement', he argued that it was 'a subject of international propaganda value which should be made without delay'.[51] Production began after an MoI meeting in January 1942 agreed that Rotha's friend, Eric Knight, should write a script and that Rotha produce the film. Knight, a Yorkshireman who had settled in America, was a film critic, novelist and sometime farmer, who is now mainly recalled for the novel *Lassie Come Home* (1942), set in Yorkshire during the Depression. Knight's most significant contemporary critical success was *This Above All* (1941), in which he 'tried to raise all the problems concerning war that face the young man of today, brought up on "no more war" and pre-Munich *laissez-faire*' (Knight 1952: 178). This can be seen as part of his passionate concern about relations between Britain and America in wartime. He came to London as a journalist in October 1941 to investigate the state of the British diet under wartime rationing and price controls, with the aim of producing articles and speeches in the interests of Anglo-American understanding. This was how he was available for the early stages of the film. The intended American distribution of 'The Strategy of Food' fitted Knight's interests well.[52] By the time that he left for America in early February 1942 he had submitted a script to the Ministry.[53] He was to continue to collaborate after his return to America, where he would be in charge of filming, and voice 'Man-in-the-Street'. From April 1942 he was working with Frank Capra for the Film Unit of the United States Army, on the *Why We Fight* series. He was killed in an air crash in January 1943, four months before *World of Plenty* was seen in public (see Anon. 1943a: 4; Anon. 1943b: 6).

Knight had corresponded regularly with Rotha after he read *The Film Till Now* in 1932. The letters range broadly and reveal a close agreement between the two men.[54] Knight's concern for the state of Britain in the 1930s led him to make a visit at the beginning of 1938, during which he had made an extensive tour of the depressed areas, and wrote some journalism. The book *Now Pray We for our Country* (1940) came out of his experiences. His interest was sufficiently detailed that he knew about the British Medical Association's nutrition report (see 1952: 86). A keen supporter of Roosevelt, Knight advocated active citizenship in a particularly vivid way in a letter written on the eve of war:

> I still believe that kindness and consideration of fellow-men is the only true interpretation of liberalism; that belief in kindness and consideration cannot just come from belonging to a party. It can only come from within each man himself – and it will spring best from him if he is a party of one man. We do not need one big party of ten million men. We need ten million parties of one man each – each one brave enough to convince himself without support. (1952: 136–7)

Rotha also refused to attach himself to any party. Knight's letters resound with themes that later featured in *World of Plenty*, including, as early as 1933, the policy of destroying crops to maintain prices (see 1952: 22, 59). The film's central theme of fair food distribution is also found in a letter from August 1941:

> Why should not refrigerator ships carry the fresh vegetables of Florida, the citrus fruits of Texas and Florida, the fruits of Pennsylvania and New Jersey, the enormous glut surpluses that rot here, to the children of the Rhondda Valley and the towns of the Tyneside? Why in the name of God not? If we can do it in war – why shall we not do these things even more so in peace?' (1952: 185–6)

Arthur Calder-Marshall was the official at the Ministry of Information (appointed in 1942) who, day-to-day, handled the detailed official side of the film's production. An author and journalist, he provides an example of how liberal-to-left figures were co-opted into government during the war.[55]

Knight, Rotha and Orr were all staying at Clifford's Inn at the time, and were expected to discuss the film 'in detail in the evenings'.[56] Orr had previously been involved with the two documentary film projects, *Grasslands of the Empire* and *Enough to Eat?*, but it is impossible to understand his role in *World of Plenty* without reflecting on his status during the war. He had intended to retire from the Rowett in 1940 at the age of 60, but with the coming of war he turned to the task of converting the calls for a scientific food policy with which he had been associated in the 1930s, into wartime food policy. He recounts that he persuaded Woolton to establish an expert committee to advise on fluctuating food supplies and their relation to the ration (see Orr 1966: 120). He built on his interwar status as unofficial leader of the loose coalition campaigning on nutritional issues but, after the demise by 1942 of the War Cabinet's Scientific Food Committee on which he had sat, Orr had little direct access to central governmental decision-making on food or nutrition. He was an establishment outsider, lecturing, publishing and pamphleteering, turning his attention increasingly to the subject of postwar food production and distribution (see Orr 1943; Smith 1992). His involvement in *World of Plenty* can be seen as part of this propagandising activity; the evidence is that he took the task very seriously.

Rotha began discussing the project with Otto Neurath in February 1942 on the basis of Knight's original script and devoted a day to detailed discussions with him in March; the first revised script was sent to him within days of its completion, in early April.[57] There was much in Neurath's interests that resonated with Rotha. We may note that in *Modern Man in the Making*,

Neurath had also referred to the adoption of scientific planning to avoid the irrationality of crop destruction (1939: 58).

Rotha, Knight and Marshall discussed Knight's draft script at the end of January 1942 and, soon after, Knight returned to America. Before Rotha submitted the first amended version in March he visited Orr at the Rowett, where they worked together to a sufficient extent for Rotha to budget for a 'typist for Sir John Orr'.[58] Orr gave Rotha his *Fighting for What?* book in manuscript; Rotha recorded that 'he is the only man I have met who has practical plans for social plans when the war ends – not the fancy architectural kind – but the sensible human security kind that deal with human needs'.[59] Orr continued to contribute throughout the production: Rotha wrote to Marshall in October that Orr's additions to the script required more Isotypes, and therefore expense.[60] Orr visited Knight twice on his trip to America in November (see Knight 1952: 218–19). Rotha was also quick to use Orr's status in promoting the film.[61]

As Rotha finalised the script, he once again turned to his wider circle, including Huxley, Ritchie Calder, Stafford Cripps and the actor and playwright Miles Malleson.[62] The archive provides no evidence of their responses, which may well have been verbal. All the same, even if Rotha's courting of influential individuals only amounts to the continued enrolment of support, it is significant which individuals he was choosing. The choice of Cripps is indicative of the political associations of Rotha's technocratic modernism.[63] Many on the left, of whom Rotha was one, saw Cripps – at this stage in early 1943 back in parliament as Lord Privy Seal and leader of the House of Commons – as a candidate to replace Churchill as Prime Minister (see Clarke 2002). In the 1930s, Cripps had gained a reputation for being an ascetic and intellectual member of the left wing of the Labour Party, a focus for technocratic and intellectually-inclined socialists. Founder of *Tribune*, he travelled far enough to the left in the 1930s to be expelled from the Party in January 1939. In 1936, when rearmament had hit the headlines, Cripps had put up most of the cash for Rotha's film *The Peace of Britain*, an abrupt work of agitprop urging people to complain to their MPs against the policy (see Rotha 1973: 164–5).

If *World of Plenty* can be seen to occupy the common ground of the core participants and many interests beyond, then the conditions of its production meant that it also had to comply with and promote the policies of several government ministries.[64] Rotha may have believed that officials saw it as a routine film about food rationing. There is, however, every sign of a high level of commitment to it in several government departments. In the Films Division of the Ministry of Information it came to be viewed as 'a very important film', [65] a fact which was used to justify greater length than originally envisaged and extra expenditure on animated diagrams (mainly Isotypes) and music. Furthermore, Rotha's stock with this Ministry was high: when he overspent on the Isotypes, Elton wrote to Marshall:

> There is little point in our saying that Rotha must bear the extra expenses himself, because he probably can't afford to do so, and, since he works almost exclusively for us and is a very important part of our production machine, we cannot allow a production company to collapse for this kind of reason.[66]

The main difficulties with the Ministry of Information focused on the American side of the production, where Sidney Bernstein, who had moved to the British Information Services office in the USA, wrote to Beddington in the MoI Films Division. He was nervous that 'to make reference to an unfortunate period in the past when crops were destroyed, and food ploughed under' betrayed a lack of timeliness.[67] Beddington told him to have this part of the script rewritten in collaboration with Knight. Rotha argued that the American government was not accused of destroying surpluses; the script merely stated it happened, not who was responsible.[68] Although amendments were made, the sentiment is still strong in the released version:

> East-Sider from New York: We are hungry.
> Second Announcer: They are burning coffee by the bushel!
> Mid-Westerner: We are tired and thirsty.
> Second Announcer: They are ploughing in the wheat crop!
> Man-in-the-Street: This is tragic! Can't someone *do* something?
> Second Announcer: They are throwing back fish into the sea!
> A Negro (loudly): Why, in God's name, why?
> Second Announcer (disdainfully): Because *you* haven't got enough money, not enough to make it worth *their* while to feed *you.*

Consultation amongst ministries was the normal mode for MoI productions; in this case, discussions took place with the Board of Education and the Ministries of Health, Food and Agriculture.[69] Only the latter two were significantly involved. The degree of attention that the Ministry of Agriculture gave to the script, compared to the other ministries, is significant. This may, to some extent, be explained on the grounds of the different cultures of, and rivalry between, the long-established Ministry of Agriculture (responsible for production) and the new wartime creation, the Ministry of Food (for distribution and nutrition), which had assumed some of the nominal responsibilities of the other.[70] The lack of contention in the Ministry of Food correspondence shows that the dominant theme of food *distribution* in

World of Plenty served their interests very well.[71] There is an implication that the Ministry of Agriculture believed its authority in agricultural matters to be challenged by an alternative notion of expertise deriving from nutrition science, in the person of John Orr. To link production to consumer nutrition, an important facet of Orr's (and Huxley's) arguments, made an explicit challenge to government and especially to the farming community that the Ministry of Agriculture administered. This is clear in the quotation from Orr's speech at Science and World Order, stressing nutritional needs at the expense of farmers' profits (see above).

The Ministry of Agriculture's suggestions on the script led to substantial alterations. First, Arthur Manktelow, one of its Assistant Secretaries, responding to the second draft of the script (written in March), complained about 'part three, where it seems to me that undue prominence is given to unofficial speakers on the subject of the future food situation in Europe'.[72] This can only be a reference to Orr, who becomes the voice of the film in the final section. Second, having failed to expand references to agricultural production to balance those on distribution, they asked for them to be cut to a minimum.

In April 1943, a series of screenings of the final film was arranged. The response from the Ministry of Food was very positive; Lord Woolton was arranging for his London staff to see the film in relays of 200.[73] The Minister of Agriculture, Robert Hudson, was also reported to be 'very impressed by the film and said that it was of the greatest importance, and that, subject to a policy point in the last reel, he would like it shown in this country and everywhere abroad'.[74] Once again, this struck at Orr's contribution, particularly the use of the example of wheat for the world food plan and his 'revolutionary' speech, with its emphasis on freedom from want. The arguments continued to within a week of the start of the Hot Springs conference in Virginia in May 1943, where it had been accepted that the film would be premiered. Elton reported to Beddington that a compromise between the 'propaganda needs of the film and the desires of the Minister of Agriculture' could not be found and so they 'agreed to delete everything which appears to be controversial', namely the use of the wheat example and 'any suggestion that production has been solved, leaving only the distribution problem, or that world trade must, as a matter of course, be based on world food'.[75] We may note the difference of view between the Ministry's concerns and the cheerful optimism of the film that 'science has the answer'.

The remainder of the discussion focused on Orr's role at the end of the film. The Minister of Agriculture maintained that 'an important man in the political or economic field should sum up the message of the film'.[76] In a clear example of clash over types of expertise it was reported that 'Lord Woolton differs from Mr Hudson in that he specifically stated that it was a very good thing that Sir John Orr came in at the end of the film before Wallace, because Sir John Orr is a scientist of world repute and not a politician'. At this stage the Ministry of Information ceased trying to satisfy any further the Ministry of Agriculture. But a very senior official at the MoI commented colourfully that 'whatever the Ministry of Food may think about it, my own view is that the film is dangerous, and all the more so because of its undoubted technical excellence'.[77]

The final form of *World of Plenty* was made possible by sufficient agreement existing between the main participants in its production, tempered by the concerns of powerful interests, especially the Ministry of Agriculture. This film shows how both the nutrition and the documentary project were bound up in hopes for the globalisation of scientific planning for human needs and a linked participatory citizenship resting on effective public communication. We can also see it as having a place in longer narratives of the establishment of nutrition science. If we were to trace a twenty-year story from 1929 we could show how, as the concerns of John Orr, one very particular nutritionist, developed from the mineral content of pasture to world feeding, the concerns of documentarists, along with various fellow travellers, also changed from local to global concerns, as with the BA. Orr made a substantial investment in the power of documentaries – his career was punctuated by a series of film projects – and his vindication came when *World of Plenty* was first shown in public, at the Hot Springs United Nations Conference on Food and Agriculture held in May 1943. There, according to a telegram sent to the Ministry of Information, *World of Plenty* 'was received with prolonged applause and excited much comment and enthusiasm'. The film allowed Orr to be present by proxy at the culmination of the conference from which he believed he had been excluded (see Orr 1966: 160).

Planning for public health

The example of *World of Plenty* – grandiose, significant, made for the state and, because of that, well served by surviving rich historical archives – can also help us to understand some of the smaller, seemingly more pragmatic, films produced by Rotha's company in this period. Looking across his output we can see the substantial extent to which he managed to live up to his initial intentions of 'producing films about progress in the fields of education, health, medicine, housing and the social sciences in terms of both war and postwar problems'. Of around thirty films and 18 issues of the *Worker and Warfront* newsreels completed between 1941 and the end of the war, the majority fall within these terms.[78]

Referring to both the large or smaller-scale films, Rotha began to speak of 'planning for public health' as a category encompassing an important part of his output. And, looking at the Ministry's wartime medical, public health and planning documentaries, we can see that his companies were responsible for a significant proportion. Paul Rotha Productions was, for example, awarded the contract for the film *Defeat Diphtheria* in 1941 and its successor *Defeat Tuberculosis* in the following year. Other films in this category included the rehabilitation film *Life Begins Again* (directed by Donald Alexander, 1942), the blood transfusion film (in long and short versions, 1942) and one on rheumatism.[79] Outwardly conventional items of health propaganda, these films were intended to embody some of the aspirational ideology on which his company was trading.

The 11-minute film *Defeat Diphtheria*, a key element in the Ministry of Health's publicity for their new national campaign against the disease, was commissioned by the MoI in February 1941 and was finished by June.[80] This was a typical product of Rotha's Unit. It was directed by Bladon Peake from a script by Donald Alexander with its commentary spoken by John Hilton.[81] Whereas *World of Plenty* represented war aims promoted by experts using an elaborate and dramatic style, this film comes across as a calm, confident assertion of one aspect of that expertise, the medical. Using mainly Hilton's voice and no other sounds or music, it is closer to *Enough to Eat?* in its style than to the larger film. Similarly, its image track is comprised both of literal shots – of children, doctors, scientists and technicians in laboratories, scenes of everyday life – and some explanatory animated drawings by Isabel Alexander. In the first two-and-a-half-minutes of the film, Hilton explains diphtheria and its common course. Like *Enough to Eat?* it plunges straight in, imparting general factual information:

> Every year tens of thousands of mothers in Britain see the doctor bend over their child and diagnose diphtheria. They watch him take a swab from the child's throat which will help him to find out if it really is diphtheria. They listen to his instructions that everything which has been near the child is to be disinfected. Every year, tens of thousands of mothers in Britain see their child taken away to fever hospital. Every year, three thousand children do not come back.[82]

The impersonal tone is then employed to discuss the nature of the bacterium, its mode of infection and the symptoms and treatment of the disease. It explains how the body naturally makes antitoxin against the infection. The next section, in a prelude to the argument for immunisation, starts by introducing natural immunity and how it is gained: 'Instead of waiting until we

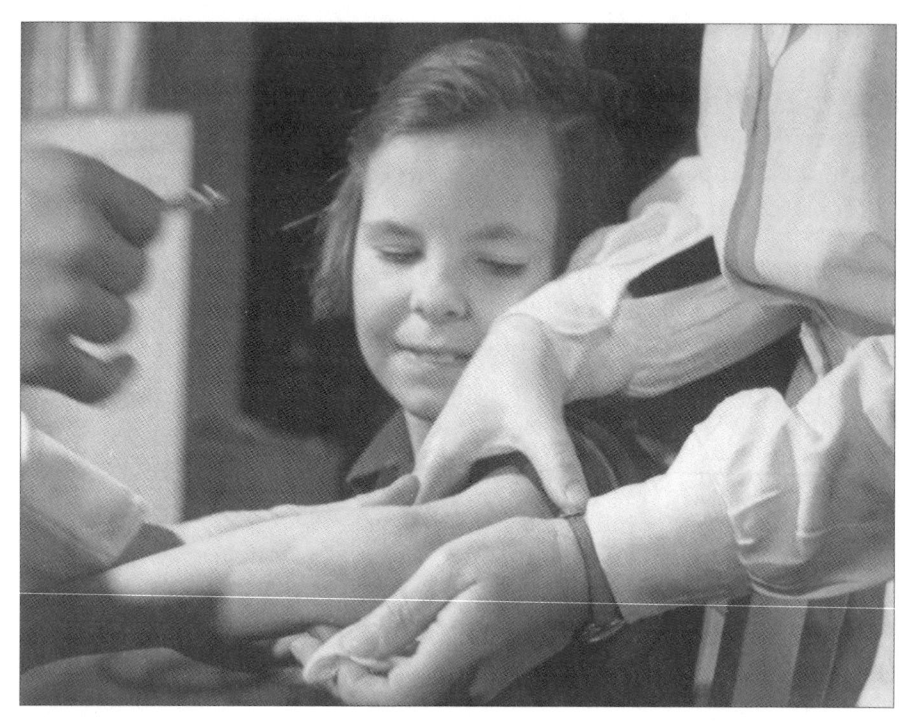

Fig 4.5 *Defeat Diphtheria*: smiling child before vaccination

get dangerously ill … it is obviously much better for us to get our bodies into the habit of making antitoxin.' The vulnerability of children is emphasised. A historical scene of Von Behring's work on diphtheria antitoxin in 1893 is followed by allusion to the development of the technique of producing immunity by injecting a toxoid (a manufactured harmless version of the toxin). Commencing with the phrase 'listen to what an expert is saying', we see a lecturer explaining to medical students the success of overseas immunisation campaigns. The British campaign is introduced with the manufacture of toxoid, and local arrangements for actual immunisation by medical officers of health and in schools is explained. A child at school is shown smilingly accepting a diphtheria injection. It is explained that parental consent is required: 'But none of these discoveries are any use if mothers and fathers do not help. Their permission must be given before their children can be immunised against diphtheria. And, if every child is made safe, then there will be no more diphtheria in the country.' In the final section of the film, its address moves away from its impersonal, fact-imparting third-person address to the direct second-person address: 'If you talk to your doctor, he will tell you how urgent it is. He will tell you that if you permit your children to be immunised, you can fight diphtheria before they catch it. He will tell you that diphtheria may break out at any time because children have not been immunised.' The change in modes of address encodes an idea of citizenship in which each parent is expected to understand and then to make the only available rational decision within the parameters laid out before them; to accept the argument of experts and the Ministry of Health and have their children immunised.

Defeat Tuberculosis, for which Rotha was associate producer, was directed by Hans Nieter for his company Seven League, which traded from Rotha's address. Nieter had been approached to produce a script in April 1942; the film emerged after many delays in the autumn of 1943 (see Boon 2008).[83] It starts with a historical vignette and then is mainly composed of a slight documentary drama narrative about two sisters in wartime, one of whom has TB and ends up being treated in a tuberculosis sanatorium. The film uses several Isotype sequences. In the first script, the film was to have concluded with the sanatorium doctor turning from the fictional characters to address the audience directly: 'Tuberculosis can attack any of you, wherever you come from, or whoever you are ... it may attack you Mrs Jones – or you Mr Smith ... and you too Mary Brown ... you too, George.' Expressing postwar themes, the doctor's script invokes joint, citizenly, action: 'It's *our* business to see that we stamp out tuberculosis ... we must stamp out slums, and poverty, and overcrowding – and we must see to it ... that never again will we allow conditions that encourage the spread of infection ... that's one of our peace aims'.[84] Although neither the conceptual trickery of changing mode of address nor the spirited speech about war aims survived into the final version released in 1943, the attempt to nuance the treatment this way shows how powerfully war aims and planning inflected how Rotha and his associates sought to represent the world.

Land of Promise

Land of Promise (1946) was Rotha's second major film on planning and postwar reconstruction. This film exemplifies the notion of a developing genre conceived with *Science in War*, and first executed in *World of Plenty*. In terms of its subject, it shows how interests and associations dating back to the mid-1930s, which were particularly resonant in 1940–41, had a continuing impact both on his own worldview, and on that which he represented to the public in his films.[85]

In spring 1941, Rotha – as ever, pursuing opportunities to make films on his preferred themes – had been in contact with John Reith, the Minister of Works, who had asked him to write a memorandum on the potential use of films for reconstruction propaganda.[86] In August, in the midst of developing ideas for *Science and War*, Rotha had begun thinking about a film on a broad canvas, touching on planning in general.[87] Collaboration was suggested with Thomas Sharp, the author of the 1940 Pelican *Town Planning*, then working for Reith, who had already submitted a script to the 1940 Council film committee.[88] In May 1942, the same month as Orr gave him the manuscript copy of his book, Rotha was considering postwar planning, and in July there was talk of him being commissioned by the gas industry to do a planning film, the idea having been sold to them largely on Sharp's script.[89] But in this period, Rotha's views on planning, reflecting Orr's influence, were broadening, as he wrote in July, also echoing Carter's aims for the 1940 Council:

> So many of the would-be planners are people who think only in terms of big-scale bricks-and-mortar planning. Rebuild our cities and roads, they scream, and disease and poverty will vanish. This represents a most dangerous slogan. I have a particular slogan at the moment – 'Planning for Public Health' – which is the theme behind these medical pictures. These countries cannot have some aesthetic-architectural plan superimposed on them from above without proper research into what human needs require. Orr is very sound on this in his 'wee bookie'...[90]

Rotha commented: 'If I do it, Sharp's script must gradually be discarded + a real planning script substituted.'[91] As we have seen, 'what human needs require' was a reference to the third of Roosevelt's 'four freedoms' explored in Orr's *Fighting for What?* Orr's argument was founded on his own expertise in food, but this book extended the principles to housing and employment because, he argued, these three were the basic needs: '*Food is the primary necessity of life* ... But the full benefits of food cannot be got until *housing* is also brought up to a health standard. A *"job"* is a psychological necessity' (1942: iv–v; emphasis in original). Like several of Rotha's associates, Orr was advocating a type of revolution but, unlike some of them, not towards a socialist state but towards a scientific one. For him the old economic system had broken down because 'it could not carry the great wealth that science has enabled us to produce ... When everyone has the necessities of life on a health standard, we shall have struggled out of the mess of poverty and degradation and reached firm ground on which we can, in safety, consider more grandiose schemes' (1942: ix, xi). Orr argued, referring to the types of groups discussed at the beginning of this chapter, that 'the rise of so many groups of planners having no connection with government or any of the great political parties is an indication of the widespread desire for dramatic changes in the economic system. The people of Great Britain will no longer tolerate unemployment, poverty and slums' (1942: x).

Once again, Rotha deferred to experts in producing this ambitious film:

I want to discuss it quietly with the few people who are qualified to talk about social planning for human needs – Orr, Neurath, Henry Morris [the educational reformer], Calder, David Owen, Hogben and perhaps a few others. But it has to be the most important film I have made – more important than the *Strategy of Food*, because it is even more fundamental. If the Gas people play, it can be the most important film of the war.[92]

In the midst of other projects, this film developed slowly; part of the process was the abandoning of an internationalist 'world planning' theme for a focus on physical reconstruction in Britain.[93] The film was finally commissioned in October 1943, and Rotha engaged the scriptwriter Wolfgang Wilhelm to work on a scenario, 'doing what Eric Knight did for the Food film'.[94] He was joined by Ara Calder-Marshall (Arthur's wife) the following month.[95] Along with Miles Malleson, Miles Tomalin and Rotha, they were working on a final script the following summer and the detailed production of the film was under way from September 1944; Rotha was editing it by mid-December.[96]

It emerged after a long gestation as *Land of Promise* in 1946, mainly sponsored by the gas industry.[97] Like *World of Plenty*, it was a film of significant scale (the final version runs to 70 minutes), representing the big theme of the role of planning of the physical environment in relation to major public health issues in the context of war and its aftermath. It also shares with that project and *Science and War* not only a common gestation in the period of greatest influence for planners immediately after the Blitz, but also the feature of the three-part historical dialectical structure preceded by a prologue. Like *World of Plenty*, it also employed a multi-voice commentary and, once again, was composed of a mosaic of visual material from many other films mixed with new shooting and music by Alwyn. It aimed to show how the heightened level of planning needed in wartime could be extended into the postwar world to overcome the public health problems of slum housing that had been a subject for documentary films since Arthur Elton and Edgar Anstey's *Housing Problems* in 1935. It also marked the convergence of Rotha's and Neurath's commitment to social planning in its application to the design and development of housing and towns. In this sense it drew on Neurath's connections with the architects of the Bauhaus and in the Museum of Economy and Society in Vienna, and on Rotha's contacts with the architectural profession dating back a decade.

The multi-voice commentary is once again played out between two main characters, 'The Voice' – an everyman figure speaking on behalf of the audience (John Mills) and 'Mr Know-it-all' (Miles Malleson). This modification of technique from the nutrition film reverses the sympathies of the viewer; unlike 'Man-in-the-Street' in the previous film, the sceptical voice is con-

Fig 4.6 ***Land of Promise*****: John Mills as 'The Voice' – direct citizenly address**

servative, backward looking and opposed to the radical spirit of the film. These characters are supported by 'Observer' (voiced by Frederick Allen), 'History' (Herbert Lomas), the offical record of parliament 'Hansard' (Henry Hallat), 'The Woman' (Elizabeth Cowell) and 'The Housewife' (Marjorie Rhodes again). In the prologue we are also introduced to 'Isotype', here not merely a visual technique, but a character in the film. Using the logotype of the figure holding a sign, 'Isotype', voiced in slightly robotic tones,[98] states 'I am Isotype. I use symbols to make diagrams. Here is my symbol for a family: father, mother and children. Here are houses people live in. Different kinds of houses. We shall be needing builders because we shall be building houses. I am here to help. I am Isotype.'

The historical dialectic of the film mirrors that of *World of Plenty*. Its thesis, 'Homes as they were' argues that lack of planning prevented reconstruction of the housing stock in the interwar period, especially of houses for rent for workers. The antithesis, 'Homes as they are' covers the wartime era. Wartime mobilisation and planning of the economy is discussed. The playfulness of the previous film is evident here again, for example in a discussion in which the Mills and Malleson characters squabble, drawing, as it were, dotted lines on a photograph of a bombed-out area. Here is a reprise of the principles of the 1942 Uthwatt Report (initiated by Reith) into the ownership of land. At the beginning of the synthesis, 'Homes as they might be' we see, for the first time, two of the protagonists, 'Observer' and 'History', with 'Mr Know-it-all' standing round the bar in a pub. They conduct a conversation with 'The Voice' who, in living newspaper style, reveals himself to be in the audience. He proceeds to articulate three problems: opening up the land so that people can have the right sorts of houses in the right places; ensuring coordination and central planning of amenities, transport and 'the whole economy'; and affordability. 'The Voice', speaking with increasing passion, once again refers to wartime measures mobilising the Services and civilians (including scientists and technicians). Of the slums, he asserts, 'they're waiting for us down there in those occupied territories … Come on then you leaders; come on, where are you? Architects, doctors, planners, engineers, social workers. Step out into the light. It's *you* who have to plan this invasion. We're waiting for the signal now.' The screen goes dark; it is Mills stepping out of the audience into the pub to argue for the urgency of planning. He speaks direct to camera. As with the tuberculosis film he addresses the 'you' of the people in the audience. The answer to 'problem number three' is an exhortation to vote and to tell MPs and civil servants what the 'millions of us – you and me' want.

Conclusions

This chapter has shown how a very particular set of circumstances led one particular filmmaker, Paul Rotha, to produce a specific genre of film that used highly elaborate cinematic technique to argue for scientific planning for human needs, with a focus on the public health issues of nutrition and housing. The technique was a fusion of dialectical montage, multi-voice commentary and Isotype, the voice of statistical truth. But even the plainer and shorter films embodied some of the spirit, derived from Rotha's associates during the phoney war and Blitz, that suffused *Science and War*. Rotha's company was not, of course, the only one making films on these themes. On reconstruction we could cite the Cadbury-sponsored film *When We Build Again* (1943), directed for Strand by Ralph Bond, and the cycle of films on the Abercrombie plan for London, including *Proud City*, directed by Ralph Keene for Greenpark in 1946 (see Haggith 1990). Equally, there were many films on medicine and public health, including *Health in War* (Pat Jackson for GPO Film Unit, 1940) and *White Battle Front* (directed by Hans Nieter for Seven League, 1940). Members of Rotha's circle also made films under other auspices; a notable example is Huxley's and Haldane's involvement in *Man – One Family*, an exposure of the biological fallacies of Nazi race theories (directed by Ivor Montagu for Ealing Studios, 1946). Particular circumstances produce particular films, as I have argued. But Rotha's was a particularly powerful and distinctive cinematic response to those circumstances. In the years beyond World War Two, Rotha's genre and the others all had to compete for the contracts that were available. Things were never going to be the same again.

5 THE FATE OF DOCUMENTARY SCIENCE FILM GENRES SINCE 1945

> As British documentary comes out of the strenuous demands of the war effort, it is faced like all the media of expression with problems of its own. Some of these are internal ... some are trade ... some are political, matters of sponsorship and information needs; and some, ideological, where documentary finds common ground with other creative forces. – Paul Rotha (1958: 227)

For my story's participants, many of whom had spent the war years anticipating the peace, the years from 1945 proved to be immeasurably different, not just from what they hoped, but also from any more sober estimates they might have made of the opportunities for film production. Here, as before, the history of scientific documentaries was part of the history of documentaries in general, and the same factors affected the narrower as the broader case. For the makers of these films, and those with whom they had found common ground, there were adjustments to be made, buoyed-up as they were on the rhetoric of the war years. To begin with, the politics were different – Attlee's government was the first single-party government in Britain since 1931, and only the third Labour administration. Economic circumstances also pressed; the government found itself obliged to pursue its programme, linked to promises of planning, state ownership and Beveridge reforms, simultaneously with responding to the stresses of reconstruction and to economic problems serious enough to reach a crisis point, in the winter of 1947.

Meanwhile, documentary makers, Rotha amongst them, worked to realise their ambitions to provide an information service stressing the subjects they were committed to. The result of all this dislocation for the making of scientific films was complex and took several years to work its way out. But it is certain that the opportunities for filmmakers were indelibly affected by the experience of the immediate postwar years, just as the postwar Labour government is generally accepted to have shaped the political landscape for the following thirty years. In the postwar decades, scientific films for projection in non-theatrical settings continued to be made in significant numbers and for the kinds of sponsors we have already encountered. Second- and third-generation documentarists continued to find employment from commercial and industrial concerns, such as the oil companies and the newly nationalised industries, and made many distinguished films of significant merit. The United Nations also promised, and in some cases provided, opportunities for documentary makers to fulfil their ambitions to make films that addressed issues and subjects on the international scale, as much in science, technology and public health as in any other area. The main concern here is to discuss the fate of the existing science film genres in the period since 1945. The economy of production was such that by 1955 television had become the site of the most vigorous exploration in the visual representation of science, technology and medicine. Participants undoubtedly perceived a sense of transition as they experienced the decline of established modes of commissioning, and delays in access to the new opportunities.

Science filmmaking at Films of Fact, 1945–47

Rotha's perception of the fortunes of documentary after the war was certainly in terms of its decline. As he reflected a quarter-century later, 'We must have been naïve to have thought that when the Labour Party actually became a more than sizeable government in 1945 it would implement an imaginative and purposeful national information service' (1973: 281). An element of this disappointment derived from earlier high hopes. We have seen how important it had been during the war for circles of filmmakers, scientists and planners to stress action and especially 'radical reconstruction' for the postwar period (see Hasegawa 1999). In July 1942, for example, Rotha had written a memorandum for Stafford Cripps, then Lord Privy Seal, and his assistant David Owen, on the organisation of documentary in wartime and the postwar period, which promoted a National Projection Board:

> Idea is to coordinate all visual propaganda after the war under one head – for national and international purposes. This means allying films to exhibitions, displays, pamphlets, posters etc … both to serve the requirements of the Govt. department + voluntary bodies such as the Trades Unions, Co-ops, RIBA's and what have you's. A big and very ambitious plan, but capable of achievement.[1]

In 1945, Rotha provided another memorandum to Cripps, then President of the Board of Trade. This, amongst proposals to establish a Government Film Corporation to support British feature-film production and distribution, also proposed a role for such an organisation to support the documentary units. By such means he hoped that sponsorship of documentary films could be guaranteed by establishing a 'planned system of production and a planned system of distribution' of up to 12 feature-length documentaries each year, with support to the organisations – such as the Scientific Film Association – who might show them. The Corporation should, he argued, work with UNESCO and other UN agencies (see 1958: 269–71).

Planning for peacetime information services had begun in 1943 (see Grant 1999: 52) and it was apparent to documentary filmmakers as early as late 1944 that the wartime footing was unlikely to be continued into the postwar period. An editorial in their newsletter argued for a national information board and insisted that 'it is essential that the National Film Service established by the MoI shall be retained and further developed' (Anon. 1944: 25). But, contrary to a great deal of lobbying and after a period in which Cabinet considered reversing the earlier decision, on 1 April 1946 the Ministry of Information was closed and replaced by the Central Office of Information (COI). There had been sensitivity that, if information were continued to be run as a ministry, there would undoubtedly be accusations of party political use (see Grant 1999: 58). The COI was set up as a technical service department for the ministries to use for the production of campaign materials, with a committee of ministers and one of officials supervising liaison. There was no Minister but Herbert Morrison, President of the Council, under whom the COI was placed, took a great deal of interest in information and exerted considerable influence over how it was run (see Grant 1999: 65–7). We are not, unlike the participants, obliged to make a judgement about the appropriateness of the downgrading decision, but we should note its consequences for state support of documentaries representing science, technology and medicine to public audiences. The postwar organisation of information services did not match the ambitions of established documentarists, and it became clear, as departmental requirements for information services developed, that numbers of commissions coming to first-generation documentarists were consistently lower than they had become used to. Rotha reverted to the agitational mode he had adopted in 1940, writing memoranda and articles, holding meetings and seeking to enrol his associates in support. For example, in August 1947 he privately circulated a memorandum which argued that

> The record of Government film production since April 1st … does not measure up to past achievements nor to the demands of the moment. No major film, comparable with those produced during the war, has been completed. Delays and obstructions have been increasingly characteristic of the commissions which the documentary units have received. (1958: 238)

The diminution of opportunities for making films and the concomitant loss of influence was dramatic. In this period Rotha did continue to make – or complete – prestige projects, notably *The World is Rich* (1947), as well as substantial films including *The Centre* (1947, directed by J. B. Holmes, about the Peckham Health Centre). Films of Fact also produced 17 issues of the newsreel *Britain Can Make It* (1945–47). Although this newsreel commission continued at a comparable rate to the wartime level, production of substantive films dropped to around a quarter. A total of five newly-commissioned films reached completion, although this period also saw the release of two delayed major works, *Land of Promise* (released in April 1946) and *A City Speaks* (1947; see Johnson 2005: 124).[2] Rotha's agitation, creating opportunities to make the kinds of films he wanted to make, included the building of relationships and the seeking of support. But the line of Rotha's company was that 'the main function and raison d'être of the documentary movement has been the public interest and that to that end the company should help the Government's information services to the fullest extent required. This being so, the Company has been geared to between 80% and 90% to COI work.'[3] He might have looked to other sponsors but he did not to any great extent, for this reason.

Britain Can Make It

The official newsreel *Britain Can Make It* was conceived to be shown under the Ministry of Information's non-theatrical film distribution scheme. It was the successor to Rotha's wartime newsreel, *Worker and Warfront*, and designed to be a part of every MoI programme. In style and technique, the new reel was completely within the genre established by its wartime predecessor; opening titles featured people at work to a signature tune, by Alwyn, in march time; as before there were nearly always three items; once again, it was weighted towards scientific and technical subjects; every issue featured some innovation in manufacture, such as aluminium houses (issue 3) and items on applied science, for example in the shape of navigational radar ('science at the helm', issue 6), or medical research, represented by the Common Cold Research Unit (issue 17).[4] As with the previous reel other companies made items that were combined by Rotha's unit, with Duncan Ross acting as his assistant producer. A contemporary document describes its ethos and this range of subjects leaning heavily towards science, technology and medicine:

> 'BRITAIN CAN MAKE IT' will be issued monthly and will carry the story of the British people gearing themselves to the task of converting a nation mobilised for war into a country constructing for peace. Above all, it will

be a pictorial of the people. Its subjects will come from industry, from Agriculture, from the Fields of Health, Science and Social Service.[5]

This is remarkably close to the ethos Rotha had chosen for his company in 1941, to produce 'films about progress in the fields of education, health, medicine, housing and the social sciences in terms of both war and postwar problems' (see chapter four). The title *Britain Can Make It*[6] may well have been a pun on *Britain Can Take It*, the documentary made by Harry Watt and Humphrey Jennings in 1940. If Britons could *take* it in 1940, what were they supposed to be *making* in 1946? This was more than a play on words; it called on a nationalistic language that represented the British people in terms of pluck and undemonstrative stamina (see Dodd 1986: 19, 21). This had been a strong feature of much wartime propaganda, and here we see it sustained into the postwar period in the service of inventiveness and productivity in the face of anticipated austerity.

The example of the first ten-minute issue shows the extent to which it continued Rotha's wartime concerns. It was commissioned by the Ministry of Information and issued in January 1946.[7] Each of its three constituent stories carries the spirit of 'making it' in a slightly different way: Admiralty floating docks were a wartime innovation, related to the 'Mulberry' floating harbour developed for the Normandy Landings.[8] This is 'Britain Can Make It' in the sense that the nation – led by inventive experts – is represented as being capable of coping with postwar technical needs. The third story, on the War Artists' exhibition, represents the British people to themselves as reflective about the experience of war. This is a continuation of the good citizenship themes of prewar and wartime documentary, in which audiences were addressed in a language which implied mature and responsible engagement with the state of the nation and of the world.

But it is the second item – on motion study – which, in this reel, most clearly reproduces the modernistic, scientistic stress of Rotha's wartime genre. This workplace technique was of a kind with Rotha's commitment to rationalistic planning solutions to social issues. Anne G. Shaw, doyenne of British motion study experts, provides a contemporary definition:

> Motion study is the investigation and measurement of the movements involved in the performance of any piece of work; their subsequent improvements, and the application of easier and more productive methods. The study of the needs and problems of the operator is the starting point of any motion study investigation as its final purpose is to enable him to work with minimum effort and maximum efficiency. (1952: 1)

Fig 5.1 *Britain Can Make It*: Motion Study – 'our old friend the left hand idling away again'

The item starts by presenting the viewer with a standard psychological assessment, a pegboard test, in which a person is asked to place steel pegs, in the most efficient fashion, into a grid of holes. An intuitive and then a rational approach to the task are illustrated. The lesson is transferred to the shop floor where the example of finishing holders for radio valves, and their packaging, are shown; in both cases how the job used to be done is contrasted with the greater productivity and ease produced by applying the science of motion study. In both cases 'our old friend the left hand' is shown to be 'idling away again, while the right hand does all the work'. In the improved procedures both hands are hard at work. The item ends with a statement from Stafford Cripps, President of the Board of Trade. First, in medium close-up, he stresses the importance of productivity to the economy, then he moves on to the impact on workers:

> We've just got to increase our production in every one of our factories as much as is humanly possible; and we must do it the easy way. 'Motion Study' is not a means of chiselling at the earnings of the workers by showing that a bigger output needs a shorter time. Its objective is the fair and sensible one of making sure that when there is a job of work to do, it's done in the way that means the least effort – and so tiredness – for those who are working on it. [Cut to screen-filling close-up, as used with Orr in *World of Plenty*.] I have taken a great deal of trouble to go into the matter from the workers' point of view and I'm convinced that 'Motion Study' will provide a solution not only for the man who has to plan the job and get it done but for everybody who has to take a part in that work.

Celebration of scientific technique had been a hallmark of interwar documentary. From 1939, this was the spirit evoked to win the war, against the odds. This is also seen with Shaw, who asserted she believed that 'by assisting in the improvement of efficiency, motion study is helping to make more secure the material foundations on which social progress rests' (1952: xii). Here, in 1946, it was being offered, in the shadow of war, as a key means to survive the peace. Anson Rabinbach has argued that concentration on the *problem* of labour was a core aspect of modernity. As he explains, within the analysis of labour, fatigue had become a key term. Also he speaks of 'an anxiety of limits – the fear that the body and psyche were circumscribed by fatigue and thus could not withstand the demands of modernity' (1990: 12). This is the sense in which motion study, targeted jointly to productivity and to the worker's fatigue, is emblematic of modernity. The motion study shown in this item is from a British tradition which distanced itself from the more familiar Taylorist 'time and motion' study (despite the fact that its pioneers, Frank and Lillian Gilbreth, had worked with Frederick Taylor). In Britain at least, the rhetoric of motion study advocates was against the 'speeding up' and exploitation inherent in Taylorism. In this item, this view of motion study is exemplified by the final sequence with Cripps professing to see it 'from the workers' point of view'.[9] The selection of the Motion Study story for *Britain Can Make It* denotes the postwar Labour government's interest in this technique, which was typical of its brand of technocratic modernism, essentially in the same tradition as Rotha's. It can be seen as exemplary of the promises in the 1945 manifesto to 'strengthen links between research and industry' (Tomlinson 1992: 75).

This newsreel story represents the people to themselves as rational actors who will accept the evidence of rationalised work, much as we saw that parents were expected to accept the argument of *Defeat Diphtheria* and have their children immunised. They are predicated as good citizens who will want their 'old friend, the left hand' not to be 'idling away', but doing its bit for the productivity drive.

The presence of Cripps in this issue has an additional resonance for this account. Rotha continued to believe that Cripps was the only member of Attlee's government who was supportive of documentary filmmaking (see 1973: 281). Both at the Board of Trade and, from 1948, at the Treasury, Cripps had an interest in the film business. He had been present in early meetings of Cabinet on whether to continue the MoI.[10] In 1946 he set up a selection committee to assist independent producers in gaining distribution of films (see Rotha 1958: 151). In 1948, we find him writing a short essay on documentary for Peter Noble's *Film Review* (see Cripps 1948). Rotha looked to Cripps and approved of his technocratic style of politics; this inspired their contacts, not only over *Britain Can Make It*, but also Rotha's one-reeler on the export trade, *The Balance* (1947), which also ends with a piece to camera from Cripps.

The World is Rich: 'A courageous, militant and disturbing film'

Postwar information services had started on a footing congruent with documentarists' ambitions, with commissions 'at the request of the Prime Minister's Office' for three major documentaries in March 1946, to cover three 'immediate publicity problems'.[11] One, on world trade, was developed by the Realist Film Unit and a second, *Prosperity Campaign*, on the production drive, was with the Crown Film Unit.[12] The third, on the 'world food situation', was given to Rotha's company. The first two were abandoned after 17 months (and an expenditure of almost £6,000) and the third emerged as *The World is Rich* in 1947. The key personnel on this film had all been involved with *World of Plenty*, but circumstances for all of them had changed. Arthur Calder-Marshall, the film's main author, had spent some time with the United Nations in Yugoslavia and had written half a dozen scripts since working as the MoI's editor on the earlier film.[13] John Orr had become the first director of the United Nations Food and Agriculture Organisation (FAO) in October 1945 (Anon. 1947c: 3). Although his commitments there prevented him from being as closely involved in *The World is Rich* as in the earlier film, the collaboration of Ritchie Calder, Orr's associate since the 1930s, may well have provided a conduit for Orr's views.[14] It is as Director of the FAO that we encounter Orr, promoting his world food plan in *The World is Rich.*

Rotha's company was asked to develop a treatment for the film on the MoI's penultimate day, 30 March 1946.[15] Like all COI films, it went through three main commissioning phases: for a first researched treatment on the basis of initial discussions; a shooting script (dependent on not only COI but also Treasury approval as it was estimated to cost above £2,500); and finally for production, on the basis of discussions around a budget submitted by the company.[16] For this film, these fell on 30 March and 2 July 1946[17] and 25 February 1947.[18] Calder-Marshall delivered the draft treatment in May 1946.[19] In July, source films were being gathered.[20] The extended gap between the second and third stages involved detailed negotiations; Films of Fact submitted a detailed budget on 9 October and again on 6 January 1947, along with a revised script.[21] A supplementary budget was submitted in April, partially because delays had necessitated filming Orr in Washington rather than London.[22] The Director of the Films Division, Ronald Tritton, understood the impact of the increase in bureaucracy in the administration of films, as he wrote:

> In the war it was comparatively easy. There were no Treasury turn-downs. There was no departmental sponsorship. Films Division, having had an idea for a film, could straightaway commission somebody to make it. Now

> there are many hoops to go through and Films Division is at the mercy of departmental whims and fancies, changes of policy, and so on. This, of course, reacts on the contractors who have to carry their staffs while they wait for the commissions to come through.[23]

Not only were there new financial arrangements for the employment of film producers, but the complexity of this particular film exacerbated financial difficulties and ensured its production was extended over 18 months. A high target price of £25,000 (*World of Plenty* had cost £6,000[24]), which officials regretted setting, had been agreed at the outset, and this necessitated arguments and fine checking at each of the three stages of payment.[25]

A rough cut was viewed by Tritton and Helen De Mouilpied, the official in charge of the production (who had been on the gas industry side of documentary production in the 1930s), in May 1947; they wrote to Rotha, proposing cuts. Tritton argued that the film was too long, 'cluttered up with too many examples and digressions' and expressed his feeling that, 'as in *Land of Promise*, you overplay your hand and become slightly strident and overemphatic'.[26] De Mouilpied spoke for the potential viewer:

> It isn't that there are not sequences where a visual repetition and the way you have cut the picture do not build up to a dramatic effect. They do, but however important the subject, I don't think an ordinary audience can take what really amounts to a build up of such climaxes.[27]

Rotha responded breezily that he agreed with much that they said, but argued that it could only be cut when commentary and music were recorded.[28] In the end, the long process over the ensuing ten months resulted in the release of a simplified and shortened version of the film.

The World is Rich was also originally intended, like the exemplar film *Science and War*, to have a prologue followed by a three-part historical dialectical structure:

> Part One: The World in Famine, as it is today.
> Part Two: The World in Subsistence, as it will be in 2 or 3 years time in Europe, and rather a longer time in Asia and Africa.
> Part Three: The World in Full Health, as it can be, provided that we tackle the problems of food and agriculture on a world scale.[29]

It is just possible to discern this original structure in the final film but, unlike the others of this kind, there are no titles to act as signposts. As a result, it comes across as a fluid, one-section film where, if its dialectic is at all apparent, it is not at the macro structural scale but at the level of contrasts drawn by the interplay of the speakers, and by the edited contrasts in the film sequences.

In visual terms, *The World is Rich* has three elements which mark it out as belonging to the genre established with *World of Plenty*. First, it was made up of a dense mosaic of footage, the vast majority of which was not shot specially for the film but selected carefully from existing films. It also made extensive use of Isotypes (although Otto Neurath had died in December 1945, his widow Marie continued to run the Institute). Once again, Rotha used multi-voice commentary. In the case of *The World is Rich*, however, because it was intended that the film be translated into different languages, there is less speech synchronised with vision (see Rotha 1958: 101); the two exceptions are Orr and the wartime head of the United Nations Relief and Rehabilitation Administration (UNRRA), ex-mayor of New York, Fiorello LaGuardia.

In the aural interplay of the film, according to Calder-Marshall's first treatment for it, the voice carrying the main content of the film, 'the man in the projection box', was intended to be not 'the conventional commentator as Emmett was in "World of Plenty" … He is setting out to explain the food situation today and in the near and distant future. He is addressing an audience of adults, whom he knows to be interested.' Perhaps responding to criticism of *Land of Promise*, the document continues 'but he also knows that the subject he is tackling is very complex and he does not resent being interrupted'. This is a significant change from the earlier type because this figure is not the pompous conservative 'newsreel' voice of *World of Plenty*, but the very model of benign expertise; he is 'the encyclopaedic expert with a knowledge of nutrition, agriculture, cooking, finance, economics and vital statistics'.[30] The other three main voices of the film were intended to be those of the men and women in the world audience 'who are able to interrupt the film's argument in order to put their point of view or ask the questions which trouble them'.[31] This model of omniscient narrator with interested interruptions might have ended up as a more forceful version of Rotha's first experiment with the technique in *New Worlds for Old*. Yet they intended a more developed interplay than in the 1938 film and there is some survival in the final version of the proposal that the film be carried by three archetypical characters: the mother, the farmer and the breadwinner or worker. But, in the final version, the exposition of information is much more equally spread between characters than intended in the treatment. The characters are not introduced as protagonists as they were in *Land of Promise*; we simply hear their voices. Nor is the comedy of interruption used to distinguish the speakers; unlike the earlier films, the composite narration is more of a conversation than an argument. Yet the voice labelled 'woman' in the script retains most of the speeches relating to mothers, that named 'negro' carries much of the universal farmer's concerns and 'voice two' often speaks from the point of view of the trades and professions.

As in the earlier examples of this genre the film's narration is closely argued, so much so that the introduction to the original treatment confessed, 'since the argument will involve a considerable intellectual strain on the audience, it is necessary to provide rest passages between one passage of argument and another'.[32] This is seen particularly in a spoof travelogue inserted towards the end of the film. But, in comparison with the earlier examples, the pace of the film also provides rest passages at the more microscopic scale; sequences of film, accompanied only by music, are allowed to 'breathe' in counterpoint to an altogether more sparse commentary. Rotha and Calder-Marshall placed great stress on the importance of the selection of film sequences, because they argued 'we have got to make people *feel* as well as intellectually realise the unity of mankind, bridging with the films the gaps between continent and continent, culture and culture' (emphasis in original).[33] Their plan was to cut gathered library film sequences to a 'scratch narration', designed to be replaced with a different narration that would respond to 'new metaphors' suggested by the selected library footage. (It was between these two stages that Tritton and De Mouilpied had made their comments.) Rotha later explained the process:

> [First] an agreement on the basic theme then the search for footage. This latter is really the first creative process and difficult to describe. It is so instinctive – and personal. You and your assistants collect footage from every available source. You sit and screen it for hours and days on end. And always you look for two things: first, footage of a *general* nature about various aspects of the subject; second, key shots which *symbolise* a specific point you want to make. (Quoted in Leyda 1964: 94; emphasis in original)

The film's prologue, accompanied by the light and shade of Clifton Parker's eloquent score, establishes the idea of plentiful food, before presenting the contrast of famine: 'One in every three people living on the earth today is threatened with death from hunger – or the diseases that travel in the track of famine.'[34] The first main section concludes that 'the problem facing the food experts was a world problem, going back before the war'. An Isotype sequence shows how some countries produced more than they ate of certain commodities, in this case wheat and rice. The commentary establishes a dialectic between surpluses (of some commodities in some countries, especially after good harvests) and deficits, endemic in some parts of the world: 'To the peoples of Asia ... two atomic bombs meant less than the age-old ravages of flood, pestilence and drought.' The second section accepts that grain was exported to feed the hungry in the postwar liberated countries, but states that surpluses were fed to animals and, as a result, food stocks fell below the safety level: 'That this hog might fetch top price in the meat market, it's not too much to say that, in another country, someone died.' The need to retain the 1,000-calorie per day diet for German miners is stressed as is the difficulty of doing so. 'Without bread, no work. To give bread ... that is the first task; to make good the food shortages produced by the war.' The third section introduces Orr's world food board, essentially a working-up of bureaucracy to deliver the world food plan outlined in *World of Plenty*. We see LaGuardia from UNRRA advocating it, in flowing rhetorical style. He mocks the market trading of basic commodities:

> I don't know if any of you have ever met a grain broker. Did you ever see a grain broker raising grain? When he goes to the office in the morning ... with his nice delicate little hands, he picks up the ticker – that's an effort – and he reads, 'Oh, the queen of Rumania's powdered her nose. Sell 100,000 bushels...'

The speakers agree that 'there has never been a world of plenty'. The FAO is said to propose an 'ever normal granary' system of retaining stocks of grain from good years to balance the bad. Such granaries, it argues, should be established not only in net importing and net exporting countries, but also in those that cannot afford to buy grain, thus allowing it to be purchased at the lowest prices as a reserve against famine. Issues are raised and problematised; for example, that giving poor farmers agricultural machinery is likely to put others out of work. The FAO is said to respond sensitively to that

Figs 5.2 and 5.3 *The World is Rich*: 'That this hog might fetch top price in the meat market, it's not too much to say that, in another country, someone died'

Figs 5.4 and 5.5 *The World is Rich*: dialectical exchanges – 'We know how to fight pests and disease, but every year more food is destroyed than was ever lost to the Black Market'

issue, planning to supply only the most appropriate machinery and 'to train the technicians, the teachers and the scientists, to develop agriculture and industry hand in hand'. Orr, introduced as the Director of the FAO, states that 'our aim is to ensure that every man, woman and child in the world shall have the food he needs to be really healthy'. He introduces a spoof sequence, a travelogue to the Middle East, the cradle of civilisation. 'Voice two' argues that use of the land has meant that even this 'earthly paradise' now relies on trade for basic necessities. 'Voice one' takes up the technocratic refrain:

> First-class roads are needed, leading to the ports. Up-to-date dockyards are wanted. Dams must be built across the rivers to supply water all the year round and to generate power, power for new industries, whose products will improve the whole life of the country. These people must have new schools, new universities, new technical institutes and new research stations.

Starting the film's conclusion, two voices conduct a dialectical conversation about the available technical and scientific means to overcome the problems of agriculture: 'We have a multitude of machines, quick, economical and efficient.' 'But still, most of the work on the earth's surface is done by hand.' This sequence concludes with a rousing synthesis: 'We have the knowledge and skill to fight all these battles against hunger and poverty.' 'If we have the faith and the courage, is not this the hope of the world?' A final antithesis shows malnourished children: 'These are the children of a war caused by the failure of our fathers and grandfathers to solve that one great simple problem: how to plan our world so that we all – yes, all – have enough of the foods we need to be strong and healthy.' 'Voice two' concludes: 'Whatever our race or religion, we want peace and we want plenty. If we pay the price, we can have both.' The music swells; a caption board reads 'The End' over shots of a field of wheat. Then Orr interrupts:

> No, no! Wait a minute. That's not the end. FAO exists. It has a plan, but the plan will only work if governments want it to work and the governments will want it to work if you, the people of the world, demand it shall work. Then this is not merely the end of the film. It's the beginning of a great world plan which will put common-sense into the affairs of men, a plan which will put politics and economics on the road to world prosperity and world peace.

This trick ending is similar in intention and address to John Mills' speech at the end of *Land of Promise*; it is an invocation, addressed directly to the audience member, of citizenly pressure on governments to pursue what is presented as the rational course. In the event, Orr's plea was too late. His world food plan was rejected by the major world powers in September 1947 (see Blaxter 2004). One of the reasons for national governments not to support it, as a leader in the *Times* had argued, was the risk that a World Food Board would keep high the market cost of foodstuffs so as to subsidise intervention stocks for the starving. For a country such as Britain, then the world's largest purchaser of foods on the open market, a duty to their consumers was in conflict with the freedom from want principles of the FAO (Anon. 1947a: 5). This rejection occurred well before audiences of significant size were able to see the film, which Rotha had first delivered to the COI on 19 August 1947. Its first public showing was to the United Nations Food and Agriculture Organisation conference in Geneva on 27 August.[35] In September it was shown at the Edinburgh International Festival of Documentary Films (see Anon. 1947b). A delay was introduced into its distribution because some delegates at Geneva, especially from the USA, had taken offence at details of the film's argument, particularly the strong assertion that animals were fed food that could have been distributed to hungry people and to the LaGuardia speech impugning grain brokers.[36] Rotha made the necessary changes by mid-September, making a version for the American audience.[37] When the film was not then booked for public showing he reverted to his customary guerrilla mode, writing to newspapers and enrolling influential associates in placing pressure on the COI and film distribution companies. In mid-October Rotha arranged an article in the *Sunday Pictorial* that placed substantial blame on the COI for not getting public showings organised.[38] It is impossible, and unnecessary, to make a judgement between the COI's and Rotha's opinions on this issue; by mid-October the COI were showing the film to potential distributors. But it proved difficult to interest the trade, who were complaining of a thirty per cent year-on-year drop in box-office receipts. One distributor asserted that postwar austerity meant that cinema attendances were dropping and, reusing an old cinema managers' argument against documentaries, that audiences went to the cinemas 'not to take on a lot of other people's troubles, but to try and shrug their own troubles off'.[39] Rotha also arranged for Woodrow Wyatt, then a left-wing Labour MP, to ask a question about distribution delays in the House of Commons.[40] Eventually *The World is Rich* received a limited London run in an edited version – 35 rather than 43 minutes – at two London cinemas, from 15 February 1948.[41] The distributors, British Lion, had received 101 bookings by the end of April.[42] But, if its primary aim was to support Orr's World Food Board, this 'courageous, militant and disturbing film',[43] though it may have moved a significant number of its viewers, was too late.

The Centre

Amongst the postwar output of Films of Fact was another documentary on a public health theme. The COI awarded the company the contract to make a film about an institution called the Pioneer Health Centre in Peckham. Despite its high profile in debates about the public health, this health centre had not been the subject of a complete film before this time. In the event, it was a Foreign Office scheme to represent British life abroad that led to *The Centre* being made. Minutes of the relevant Cabinet Committee explained the rationale behind such publicity for use overseas:

> The object is to spotlight those things which show Britain as a strong and vital factor in the world, and to illustrate by practical examples the distinctive contribution she can make; to give a dynamic picture of Britain in constant evolution, though firmly rooted in an unbroken tradition, branching out into new developments in politics, in social affairs, in scientific research and also in other spheres.[44]

The Peckham Health Centre, as it is generally known, was an eccentric choice to fit this brief, as it was a singular phenomenon. Not a health centre in the now familiar sense, it was a social club for families, with an added measure of medical surveillance undertaken by its founders as 'an experiment in the living structure of society', as they described it in their book title in 1943. The aim of its founders was to investigate the links between members' lives and their pattern of health, in pursuit of a definition of 'positive health' (see Pearse & Crocker 1943; Lewis & Brookes 1983). *The Centre* was written and directed by J. B. Holmes, whose career had taken him from Gaumont-British Instructional in the 1930s via the Crown Film Unit, where he has been described as one of those responsible for the development of story documentaries (see Brown & Enticknap 2003). Working with Rotha, he had produced at least two issues of *Britain Can Make It*.[45] The film starts with a short factual introduction which explains the geographical location of the Centre, and the social composition of Peckham, a residential district. It shows children arriving to play at the Centre after school, followed by adults moving 'from working to living'. The commentary, in calm, warm tones asserts:

> People don't kill time at the Centre; they use time. They acquire physical and mental health. [Shots of sports and exercises under way.] They can do what they like, when they like, as they like [shots of people in gymnasium], and in this way the Centre becomes a community of families and not just a mass of individuals. [A focus-pull reveals a man and a woman watching the exercise through a window from above; there is a cut to these two, who are shortly revealed to be the founders, Dr George Scott Williamson and Dr Innes Pearce.] But there is another side to it, the scientific side. The directors and their staff are really biologists, studying the actions and the behaviour of families. And the building has been designed with big spaces and glass partitions so that the staff can come into the closest touch with members, and watch how each family makes use of the space. The members are, in fact, co-operating in a unique piece of research into social biology.

The commentary rounds off this section by explaining that the Centre was a private enterprise, running alongside the government health scheme and supplementing it. The second, and longer, section of the film is then introduced by a female voice. This is in documentary-drama style, a technique first used at the GPO Film Unit a decade before. It involved writing an ideal-typical fictional narrative to convey the experience of participants. Here, as generally, it was acted out by non-professional actors, in this case participants in the Centre selected for the purpose. This is the story of a new family joining the experiment, Maureen and Fred Jones and their school-age son, Johnny. Using the device of a member showing them round, we are introduced to some of its facilities. The conventional documentary 'wild track' of voices is evident. Then the directors speak to the Joneses, asking their opinion. Scott Williamson gives a speech, explaining their aim, 'to study health'. The annual health screening is explained. However, Maureen Jones becomes resistant to the Centre after Dr Pearce tells her that she needs to have an operation to correct a 'misplacement' before she can have a second child. This introduces the notion of the significant amount of undiagnosed ill health in the population, one of the experiment's concerns. Scott Williamson's voice explains that their policy was not to insist on treatment, but to leave Maureen to make up her own mind. Eventually, in the story, the whole family stops attending. Maureen, in her turmoil, says she thinks the Centre's

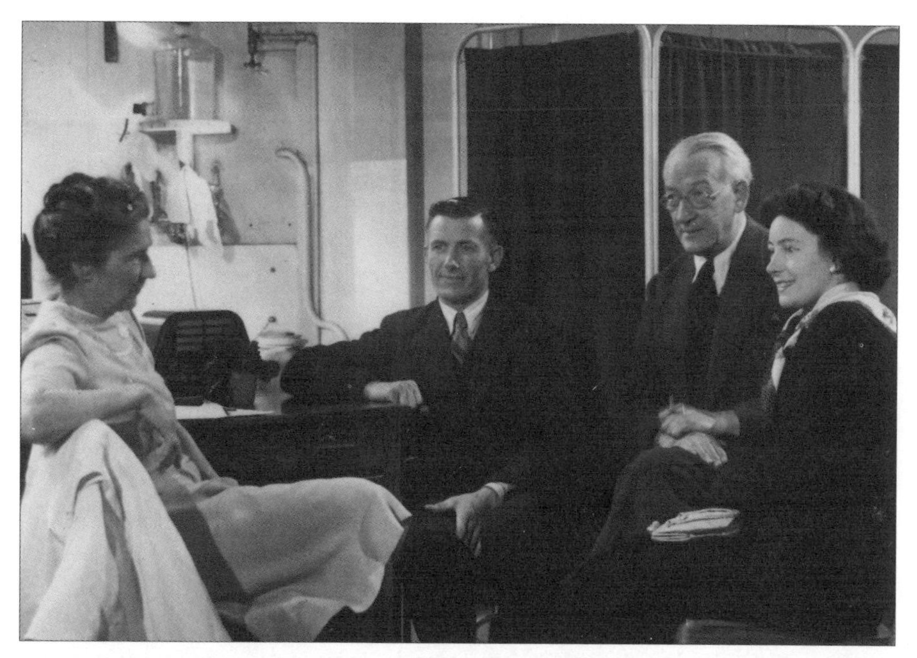

Fig 5.6 ***The Centre*****: documentary-drama – (from left) Innes Pearce, 'Fred Jones', George Scott Williamson and 'Maureen Jones'**

directors are 'cranks'. Eventually she concedes the point; she has the operation, and joins in at the Centre again. She becomes pregnant. 'This is a time when both parents need all the information that the scientists can give, and they grow more and more eager to make use of it.' The ways that the Centre can support new mothers are highlighted; the psychological benefits of the family unit and the vitality of the pregnant woman are emphasised as more and more of the Centre's facilities are shown. In the film's closing shot, the main musical theme – an arrangement of 'The Centre Waltz' composed by a member of their amateur dance band – swells. We see the family, complete with new baby daughter, being greeted by Scott Williamson and Pearce as they sit in the sunshine at the Centre.

The Centre stands for a radically different approach to making films about medical, scientific or public health themes from Rotha's major projects or his newsreels. It might be seen as a successor and enlargement of the documentary-drama strategy originally intended for *Defeat Tuberculosis*, but here the technique is fully extended and a real attempt is made to round out the characters, and to give Maureen a dilemma that might engage the viewer. Increasingly over the decades ahead, and especially at the Crown Film Unit and BBC, this genre became one of the major resources for the representation of social science and medical themes to lay audiences.

The end of Films of Fact

Paul Rotha's company went into liquidation at the end of 1947 after Humphries Film Laboratories, who were owed several thousand pounds by Films of Fact, demanded payment. It was not ideology but money that caused the end of the project launched at the start of 1941.[46] The COI had been investigating since at least September 1946 whether the state of Rotha's company finances excluded him, by Treasury rules, from being employed by government departments.[47] The conclusions of these investigations, which were being discussed even as *The World is Rich* was being dispatched to the FAO conference, undoubtedly reduced his chances of being commissioned to make new films. The archive justifies Rotha's belief that both these rules, and the delays caused by discussions between client Ministries and the COI, contributed to his financial difficulties.[48] It also shows how COI officials became increasingly discontented with Rotha and his seeming unwillingness to accept the government rules under which they and he were obliged to work. However, whilst resentfully registering his public denunciations of COI delays they continued to value his skills as a filmmaker. Robert Fraser, COI's Director General, revealed this in an internal note, during their discussions of how to respond to information on Rotha's financial position:

> Though Paul Rotha can be very tiresome + has been very unamiable + even disloyal to the Central Office, he plainly has his own problems and anxieties as a result of an obstinate + admirable determination to keep the 'social documentary' alive throughout the last 20 years. He is now one of its fathers. And though not perhaps one whose influence is entirely for its good, he has a record, a position, + importance that makes me feel we should go rather gently.[49]

We should note that the model of social documentary that Rotha stood for, and which Fraser would have recognised, stressed the force of rationality and science in creating freedom from want.

Rotha's agitation, partially via the Federation of Documentary Film Units, led to an investigation by the Cabinet Office into relations between the Documentary Units and the COI. This provoked a series of exchanges that reveal how COI staff were coming to think of the documentarists. One report, sent by Tritton to his superior B. C. Sendall (Controller of the COI's Home Department), having listed directors of wartime feature documentaries and documentarists who had moved into features, stated:

> Left behind is the old Soho Square gang – the hard core of documentary: Anstey, Wright, Rotha, Bell, Alexander, Bond, and a few others. But something has happened to them since they made documentary the British contribution to the art of the cinema fifteen years ago ... they are stale ... they lack inspiration ... they are too political in both the broad and the narrow sense ... They see themselves as a great Power for Good. They are conferring with their documentary colleagues all over the world in pompous conclave, but they aren't making good films ... They are running a closed show and are unwilling to let in the new boys with the bright ideas or even to look for them because by so doing they might prejudice their own position.[50]

Even Helen De Mouilpied, who had consistently argued Rotha's side, conceded that 'documentary has always tended to be made by the enlightened for the intelligent and it has never had wide distribution. The Third Programme is surely not the medium for national publicity.'[51] Sendall revealed the way that Rotha's public criticism – he referred to 'the Rotha faction' – of the COI had struck home: 'It is necessary to rebut emphatically the idea that the documentary film in Britain is degenerating as a result of incompetence and lack of imagination on the part of the Central Office.'[52]

Part of the documentary makers' adjustment to the postwar world parallfeled their experience in 1940. They had carried a belief, since Grierson's first EMB memoranda, in the role of documentary as a core component of the nation's information services. For much of the war, and certainly after 1941, there had been a coincidence of interests between them and the policy of the MoI. Their citizenship themes, showing people at work in modernity using technology, had worked effectively as home front propaganda; at the same time, major themes such as that of nutrition in *World of Plenty* had done effective service in international projection. But, with the establishment of the COI, they were competing as outsiders on the territory of the insiders – the new officials responsible for information services. Inevitably, tensions arose.

Although Rotha continued to direct films – and to seek to do so – for at least a decade and a half after this, *The World is Rich* was the last film on the *World of Plenty* model. The Isotype-enriched historico-dialectical multi-voice commentary film on major scientific or public health issues died with the solvency of Rotha's company. It is clear that Rotha would have continued to develop the form; for example, he used it for a 1956 script for the United Nations on the subject of atoms for peace, *The Power of Peace*, which was cancelled for financial reasons (see 1958: 101). Other filmmakers scarcely used the technique of multi-voice commentary that made these films so distinctively Rotha's.[53] Basil Wright, for example, seems to have considered it to be Rotha's creative property or signature:

> Rotha did this distinctive thing of developing his own particular style in *World of Plenty*, *The World is Rich* and *Land of Promise*, a very, very difficult and dangerous way of filmmaking because it didn't permit any form of escape for the audience. (Quoted in Pearson 1982: 69)

Without multi-voice commentary, any other documentary film, however forceful and conspicuously well made, and however much it assumed the mantle of promoting freedom from want in scientific films, cannot be considered to belong to this genre. The themes of science and society did continue in other films, and these are considered later. But *The World is Rich* was perceived as a terminus by others too: Rotha's friend, the critic Richard Winnington, in his review for the *News Chronicle*, suggested that it 'closes the didactic phase of documentary on themes of contemporary life. The individual, contrary to the Grierson tenet, becomes more and more important. He will be the core of the new documentary, whose aim will be to reach your mind not by giving a lecture but by telling a story' (Winnington 1975: 74). Despite the fact that Winnington's judgement was only partially borne out, his perception of the changing genre requirements of the age was astute.

Other 'members of the old Soho Square Gang' perceived a significant decline in filmmaking opportunities during the Attlee years. Despite the fact that COI officials were able to show that the total length of the films produced in 1946 exceeded the two previous years,[54] they also conceded that circumstances had changed considerably. Tritton commented that the task was more difficult than with the wartime films, because 'the whole country was behind the war effort and therefore interested [but] the whole country is not behind the present methods of tackling the aftermath of war problems'.[55] Films were becoming more expensive, contemporary themes were 'complex and shadowy', leading to less work, and the average length of films was creeping up, also leading to fewer commissions overall.[56] The COI began to employ younger documentary filmmakers; for example it was Data Film Productions, led by Donald Alexander which had broken from Rotha in 1944, who made the documentary-drama *Here's Health* (1948), the most extensive of the NHS films. They also broadened the cinematic techniques they adopted, as for example with the cartoons made by John Halas and Joy Batchelor used for the National Health Service and the National Insurance Campaigns (see Wildy 1986: 13). It was a little more than purely symbolic when Churchill's re-elected Conservative government closed down the Crown Film Unit, successor to the EMB and GPO units in 1952 (see Harding 2004).

Explaining the documentarists' experience of decline

Documentarists such as Rotha might well have consorted with various kinds of expert professionals in science, planning or health, but in the new polity the COI was not primarily concerned with purchasing filmmakers' expertise in anything other than filmmaking. At the same time, it cannot have helped Rotha's cause that both his associates, and the views they espoused, enjoyed mixed fortunes in the second half of the 1940s.

The advocacy of planning in Rotha's films rested on his association with like-minded individuals, including scientists, politicians, architects and planners. Just as there were competing visions of the role of information between documentarists and officials, there were discrepancies between his associates' views of planning and those developed and enacted under the auspices of the postwar Labour government. They were competing – from the outside – on the territory of insiders, the Ministers in the pro-planning postwar Labour government and their officials. The wartime advocates of planning had to take their chances as advisors and members of committees set up by the government. The clarity of the situation was not helped by the fact that there were several competing notions of both economic and physical planning, not only amongst politicians – Richard Toye speaks of 'the vague and often contradictory nature of Labour's conceptions of planning' – but also amongst officials; one of the results, he argues, was that 'despite tentative initiatives and much talk of planning, the promised planned economy showed little sign of materialising' (2000: 81). Cripps, both at the Board of Trade and at the Treasury, was drawn to the more radical 'Gos-plan' style, of planning the quantity and prices of consumer goods and the quantities of export and capital goods. He asserted, however, that to *compel* planning would be inappropriate and that it was important that planning should be willed by the electorate; 'that is why democratic planning is so very much more difficult than totalitarian planning' (quoted in Toye 2000: 92–3). We may note that this model of democratically-willed planning is very like Rotha's in *Land of Promise* and *The World is Rich*. But the failure of the planning agenda in the postwar Labour government was not only a matter of conflicting understandings, but also of the difficulties of the economic situation (see Toye 2000: 81). *Economic* planning, by this account, was a casualty of the local circumstances of the late 1940s.

Other types of planning also suffered. In town planning, for example, the aspirations of the 1940 Council and other linked groups also often failed to reach expression in steel, glass, concrete, bricks and mortar. Where 'radical reconstruction' schemes often focused on rebuilding town centres, the economy dictated that for over a decade after 1945 priority projects such as the provision of housing took precedence over planned civic and commercial amenities. But the new housing estates were also frequently built without the social facilities which planners saw as essential to lives healthy in the psychological and social sense. Larger schemes fell prey to many of the problems described in *Land of Promise*, not least a shortage of capital (which prevented compulsory land purchases, now permitted under a 1944 Act) and of skilled labour (see Hasegawa 1999: 149–61). As Junichi Hasegawa concludes, 'what is striking is how quickly ideas of radical reconstruction became forgotten both in the policy-making process and among the public' (1999: 155).

If we follow the argument that the documentarists' fate paralleled that of their associates in planning, the same case may be made in relation to the advocates of the social role of science; it might be argued that the decline in opportunities for first-generation documentarists was a matter of party politics or of political ideology. Tritton's comment that 'they are too political in both the broad and the narrow sense' would add credence to this interpretation. Some historians of the social relations of science have described an eclipse of the influence of the scientific left – including some of Rotha's associates. The balance of power amongst organised scientists shifted away from social engagement towards an emphasis on the freedom to pursue pure science, where Michael Polanyi's Society for Freedom in Science was influential (see McGucken 1984: 265–306). Gary Werskey similarly suggested that the ascendancy of the scientific left within the BA's Division had relied on a 'popular front' linking 'reformist' and 'radical' scientists. He argued that this 'front' became divided, both because reformists stepped back from activism on account of 'a gradual improvement in the government's treatment of science and scientists' (Werskey 1971: 81), as well as the shift in concerns this permitted towards issues of freedom in science. But, if we disaggregate Werskey's categories, we can see an alternative account in which a variety of individual scientists (and planners, such as Edward Carter), each with particular politics, experience, seniority and scientific expertise, was simply pursuing the new opportunities of the postwar period. This followed the 1930s and wartime pattern when they had moved in often temporary alliances, in which political motivations might often come second or third in the pursuit of other commonly-held concerns. Parallel with Werskey's history of political commitment, there is a history of interests to be written here. Like the planners, as outsiders they were obliged to take their chances in relation to the government of the day in pressing their suit. Werskey notes that the BA 'simply allowed' their Social Relations Division 'to wither away during the Fifties. What spirit of activism remained was channelled into the World Federation of Scientific Workers, "Science for Peace" and…' the Campaign for Nuclear Disarmament (CND) (1971: 81). We may choose to see this as the application of their energies to the urgent issues of the day – as they had during the war – rather than only as the failure of an earlier project to construct a socialist science within a communist state. However significant a factor this was for Bernal and Hogben amongst Rotha's associates and for the remainder discussed by Werskey (Haldane, Hyman Levy and Joseph Needham), the fact is that Rotha had been most influenced not by members of the communist left, but by more liberal figures such as Orr, Huxley and Zuckerman. Also, it is clear that many of these scientists, including Zuckerman, continued to be influential within government in shaping attitudes to, and arrangements for, science. He became a member of the Committee on Future Scientific Policy chaired from late 1945 by the Treasury civil servant, Sir Alan Barlow, whose conclusion was that an Advisory Council on Scientific Policy should be formed, essentially extending into peacetime in modified form the wartime experiment of the Cabinet Scientific Advisory Committee. Zuckerman was one of its members when it met for the first time in January 1947 (see McGucken 1984: 335).

Evidence for any absolute decline of influence for the scientific members of Rotha's circle is not abundant. Ritchie Calder, awarded a CBE for his service on the Political Warfare Executive, pursued an energetic postwar double career in science journalism and as consultant and committee member. He was science editor of the *News Chronicle* for a decade from 1945 and a member of the editorial board of the *New Statesman.* He was also a member of the Fabian Society's Executive Committee. Orr invited him to attend the FAO famine conference in 1946 as a special adviser, and he was on the delegations to the UNESCO general conferences in 1947 and 1948. With Huxley's support he was appointed to UNESCO's Mass Communication working party (see Ellis 2000: 232). He subsequently undertook several missions for UN agencies and continued to publish works of popular science (see Williams 2004). In the midst of all this activity, he was also involved in the scripting of *The World is Rich.* Later, he worked as a film commentator (*Mirror in the Sky*, 1956) and commentary writer for television (*The Unforgotten*, 1961).

In the same period, several of Rotha's other associates took up opportunities provided by the establishment of the United Nations agencies. John Orr was director-general of the Food and Agriculture Organisation between 1945 and 1948 (see Blaxter 2004). Julian Huxley successively became secretary-general of the UNESCO preparatory commission and, from 1946, its director-general, a post he held for two years (see Olby 2004). He took with him his old associate Edward Carter, who had been the leading light of the 1940 Council, to become the organisation's first librarian. He also made John Grierson director of Mass Communications and Public Information, who in his turn appointed Basil Wright to plan a series of films.[57] In part, these appointments reinforced a complex generational effect in which new opportunities opened up and new roles and behaviour were expected of those associated with the preoccupations of the prewar world, not only those nearing or beyond retirement age, such as Huxley or Orr, but also younger individuals, including Rotha.

If we look slightly further ahead, the influence of some of the individuals we have encountered continued to be significant. Hugh Gaitskell, original member of the Tots and Quots, at the same time as he was Minister of Fuel and Power in the postwar government, continued his associations with those in science concerned with its social relations, for example joining the AScW. When he became leader of the Labour Party in 1955, he established an informal scientific advisory group that soon numbered newly influential and scientific figures, such as Jacob Bronowski, Desmond Bernal, P. M. S. Blackett, C. P. Snow and Solly Zuckerman. It was in these discussions that the famously technocratic spirit of the 1960s Labour government – complete with a National Plan – was forged (see Horner 1993). Also, when Harold Wilson came to power in 1964 it was Zuckerman that he made Chief Scientific Adviser to the government (see Ziegler 2004).

In general then, we must beware making the postwar half-decade a terminus for our story. It is true that Rotha's career in documentary films faltered critically at the end of 1947, although he did continue to make films after this date, and we shall encounter him working at the BBC in the next chapter. As the postwar polity matured and the economy improved there continued to be opportunities for filmmakers to make films, not least via the COI, for the nationalised industries and for the oil companies. But most opportunities went to second- and third-generation documentarists.

Science films after 1950

It is important to balance the experience of the first-generation documentarists with the broader picture of non-fiction film production, and to note other areas where production was expanded. Not all of Rotha's contemporaries in documentary shared his long-term experience of reduced opportunities. Edgar Anstey, for example, was installed from 1949 as the head of British Transport Films (BTF), the unit established to promote public transport. Younger filmmakers, including those working for Data, acquired significant commissions including, for *Mining Review*, a newsreel from the newly nationalised coal industry from 1948. Both of these new state-funded units gave considerable scope for the representation of science, technology and health issues. The National Coal Board's newsreel, for example, started its first issue with an item on the Meco-Moore automated coal cutter; amongst many items on occupational health it covered a miners' health centre in a 1948 issue. The majority of BTF output represented the railways and the Britain that they gave access to, both promoting new ventures such as *London's Victoria Line* (1969), retelling industrial history, as in *Age of Invention* (1975) and even producing the occasional natural history film such as *Between the Tides* (1958). Furthermore, concentrating too closely on documentary may divert attention from other types of non-fiction filmmaking. *This Modern Age,* the Rank Organisation's newsreel in the *March of Time* mould, not only treated typical documentary themes, including some representing technology (*Tomorrow by Air,* 1946), public health (*Homes for All,* 1946) and nutrition (*Will Britain Go Hungry?,* 1947), but also gave temporary refuge to individuals associated with documentary, including Edgar Anstey and John Monck (see Enticknap 1999: 2). News magazines such as Rank's *Look at Life* (1959–68) and Pathé's *Pictorial* series (1918–69) also continued to represent some science, technology and medical themes.

One measure of the health of scientific filmmaking for projection (rather than broadcast) is the activity of organisations. The Scientific Film Association (SFA) continued, via several mergers, for many decades into the postwar period. From the late 1930s, they worked to classify and promote the use of all kinds of scientific films, whether for public or for specialist audiences. A publication in 1946 set out to assist those assessing such works. Amongst the 18 categories of scientific film they defined were:

> *Record.* – … registers or represents phenomena or events (such as scientific experiments) as they normally occur, and is not primarily concerned with explanation.
> *Demonstration.* – records phenomena, events or processes carried out specially for the camera.
> *Instructional.* – … an aid to acquiring knowledge …
> *Hortative.* – … one which encourages, inspires or recommends a particular course of action by appeal, mainly or partly, to the emotions.
> *Descriptive.* – … presents a group of facts in a coherent form, but without showing the group's relation to, or interaction with, groups exterior to itself …
> *Interpretative.* – … presents a group of facts with special relation to their social or economic context.
> *Factual.* – … lays emphasis upon the events with which it deals, and without reference to their relation to any particular set of characters.
> *Fictional.* – … lays emphasis upon the relations of a particular set of imagined characters to a particular set of real or imagined events. (Anon. 1946a: 15)

This publication gives us a way of understanding how contemporaries with a particular concern with scientific films thought about them. They showed how the categories could be applied using the example of Geoffrey Bell's *Transfer of Power* (1939), a historical, descriptive, interpretative, factual film. By contrast, the model we have seen developed by Rotha would certainly have been considered 'hortative' and 'interpretative'. It was not the case that the SFA excluded the hortative from their showings but, unlike Rotha, they were concerned with the whole range of scientific film genres, not just those on the model of *World of Plenty* or *Worker and Warfront.*

An International Scientific Film Association (ISFA), based on the SFA and the French Institut de Cinématographie Scientifique (active since 1933), was founded at a congress in Paris in 1947, with membership from 21 countries and UNESCO (see Anon. 1952: 3). Over the ensuing years ISFA held regular congresses, where a wide variety of films was shown and discussed. In summer 1959 they established a working party, chaired by Edgar Anstey, to promote the popularisation of science and technology by film and television (other sections were concerned with research films and educational films) (see Tosi 1960: 243). At their 1960 congress in Oxford the popular science section showed – to 1,500 members of the general public as well as delegates – nearly 100 films, totalling nearly 24 hours running time in their 'popular science festival' (see Chibnall & Le Harivel 1960). Two pre-published papers in the popular science section, by Virgilio Tosi and Igor Vassilkov, were concerned to settle 'the years-long argument on what is a popular-science film' (Vassilkov 1959: 55), their category that approximates to the concern of this book, that is, scientific films for the general public. Both speakers accepted the distinction of popular science from research and educational films, and both argued from the contemporary expansion of high technology. Tosi was more concerned with the potential to make dramatic films popularising science, despite what he saw as the poor quality of many popular science offerings. For him, the importance of science and technology to the postwar world created an opportunity for 'those who have the mind and sensibility of an artist … to interpret the vitality and drama that exist today in the scientific and technological problems which man has put to himself … and with which life and the future of humanity are invested' (Tosi 1960: 245). It was the increasing importance of applied science, displacing the older model of the 'secluded' pure scientist, and the consequent financial demands on government budgets that made science of significance to public opinion (ibid.). Vassilkov spoke of popular science films as 'spreading through the medium of poetic inspiration and artistic imagery, the spiritual culture of mankind'. In a surge of techno-enthusiasm he continued that this age of 'scientific genius' featured the scientific demolition of 'the seemingly unbreakable wall between the organic and inorganic worlds. And the utilisation of nuclear energy, the ultra-high-speed jet aircraft, the gigantic proton-synchrotrons, automation and remote control in industry, the launching of artificial earth satellites!' (1959: 54–5). His definition was that:

> A cinematographic work may be called a popular-science film in which the basic task for this kind of film is set and solved – the popularisation of scientific knowledge, where in a simple and lucid form, clear to the broad masses of spectators, the fundamentals of this or that branch of science or the new discoveries and achievements of science and engineering are expounded. (1959: 57)

This is a long way from the model of the hortative film conveying science promoting freedom from want that we saw with Rotha; it is promoting a powerful and confident science and technology, but one focused on abstract research and on applied science visible in the high technologies of the postwar period, not in the service of solving social inequality.

Increasingly over the first two postwar decades, scientific filmmaking was seen in the light of, or in comparison with, television. For example, a meeting of the Society of Film and Television Arts in 1963 discussed the coverage of science in the two media. There Geoffrey Bell, with a quarter-century of experience in scientific documentary making, devoted his talk to what he called 'the factual cinema' – non-fiction science films which, following the accepted general sense of the SFA categories, he separated from scientists' research films and features. We should note that Bell, a member of the original documentary group, continued to use conventional documentary citizenship language to justify science documentaries. He stated that 'this last kind of film is essential for a scientifically informed electorate. Without that sane progress cannot be expected from a twentieth-century nation' and commented that 'the need is just as great today to relate ordinary people with technology, if technology is to be kept the *servant*' (1963: 4; emphasis in original). He linked this to the prewar scene, arguing for a continuity of practice; that 'from the early days of the documentary movement, Grierson asserted the value of the screen as a window on science and technology. At the GPO Film Unit he could hardly do otherwise … So documentary supplied a nucleus of scientific films' (ibid.). He referred to three then-recent examples, *Food or Famine* (a scientific appraisal of the world food problem, directed by Stuart Legg for the Shell Film Unit, 1962), *The Peaceful Revolution* (the impact of electrical technology on India, 1961), and the recently televised *A Hundred Years Underground* (the growth of the London Underground, directed by John Rowdon for British Transport Films, 1963).[58] The remainder of this section will describe the first two of these films and compare them, as examples chosen by an engaged contemporary, with the genres of scientific films we have already discussed.

The Peaceful Revolution

The Peaceful Revolution was directed by Atma Ram (who had previously made films on malaria and the soil for the Shell Film Unit). Its sponsors, Associated Electrical Industries (AEI), were contractors for some of the initiatives illustrated. It was made under the conventional auspices of postwar documentary, with Film Centre providing the production and Michael Clarke producing. (Clarke had started with Rotha at Films of Fact and had been one of the Assistant Directors on *The World is Rich.*) The film's cinematography was by Wolf Suschitzky, veteran of many documentaries, including *World of Plenty*. In *The Peaceful Revolution*, which takes as its topic the transformative power of hydro-electricity on the economy and culture of India, we can see the longevity of themes of technological modernism from before the war, in this case placed into the context of international development.

This 26-minute film starts with a wordless two-minute prelude that takes us from mountains, via melting snow and streams, to a major river, accompanied by sitar and flute music composed by Hemant Kumar. We see traditional net fishing and people washing on the banks of the river. At that point the commentary begins in documentary-poetic mode: 'All reverence to you, rivers of India. At your touch the desert is brought to life, the seed bursts, the wheel turns. Your power is of nature herself.' Then the scene changes to show a dam, and the commentary builds on the contrast between the old and the new: 'Bhakra, the highest dam in Asia ... One of the temples of the new India.' The commentary celebrates the dam's impact on the irrigation of six million acres of land, doubling their yield. The film turns to its main theme, hydro-electricity, with turbines and associated equipment built by an international team. 'This is the new India, bringing together materials and experience from all over the world to tackle the massive problems of her own development. And the need of this new age is energy.' Shots of rotating generator shafts are intercut with flowing water then, in a sequence accompanied by drum and flute music, with scenes of trade (the traditional documentary trope of ship-unloading), weaving looms, electric trains and steelworks. 'More power, more heavy industry, more steel; the means to make India self-reliant in the modern world, able to build her own trains and tools and machines.' We see the documentary convention of people at work in modernity with technology, although on this occasion, of course, Indian people. 'Three steelworks – Russian, German and British built. These too are temples of the new India, birthplaces of a new industrial revolution.'

The film then introduces an antithesis; the scene changes to threshing by hand and mules at work. 'Away from the din of the factories another India waits, simply, in a way that has altered little in a thousand years.' We see ancient stone carvings, accompanied by a slow flute and sitar raga. There follows an interrogative synthesis, effectively addressing issues of freedom from want:

> Questions waiting for an answer: how can India build a society, modern and prosperous, but as much her own as the civilisations of her past? How can ancient ways of growing food be improved as the number of people increases: 438 million people and another eight million every year? How can India stand on her own in an industrial world? How can there be work and a good life for the growing numbers in cities and villages alike?

The answers forge together the technocracy of *We Live in Two Worlds* with a planning discourse not unlike that of *The World is Rich*, supplying relief from want, in a way that is familiar from Rotha's whole cycle of world films:

> These problems shadow every moment of Indian life and government. And India seeks the answer in the great national effort to plan and build. The five-year plans have a simple purpose: to spread social justice to everyone and to make India truly self-supporting. And so there must be plans to exploit to the utmost the hidden resources of the land and nature.

The film turns to consider the next generation of Indians as a resource for development, described as 'the hope of India, their minds new and agile, responsive to men's understanding of nature and so to the science that could change India. In a school or a museum, a revolution can begin.' University and polytechnic education in science and engineering is stressed, with a section on the Heavy Electricals Plant training school in Bhopal (AEI's part of the project). Learning how to work in metal rather than the traditional crafts of wood and stone is emphasised. In modernistic free flow, the commentary asserts that 'a new kind of person is emerging' and firstly we see glassblowing followed by the mechanised manufacture of light bulbs. The film shows the construction of an electrical grid for India so that 'electricity can reach out to the developing villages', powering tube wells for areas distant from rivers. We see sequences of the erection of poles for electrical supply in a village and the use of the power source for the grinding of corn, street lighting and lamps in the home, where extra hours can be applied to developing skills.

The Peaceful Revolution is a continuation of both the subject concerns and the representational forms of earlier documentaries. The technological modernism of a world transformed by the application of science and technology shown here is entirely congruent with the subject concerns of prewar films such as *The Face of Britain.* Incidental details of the representational form also borrow from the longer documentary tradition; people at work accompanied by the wild track of workmen's voices, for example. The construction of the film has the ghost of a dialectical structure, but only in the loosest of senses; traditional and rural India is compared to the potential of hydro-electricity, which, in its turn, is put up against the work that electrification must do to create 'the new India'. This goes on to provide a springboard for the account of the role of technical and scientific education in the country's modernisation. There is a single voice of commentary in the style of the documentary mainstream, rather than Rotha's late style. The employment of this educated English voice continues a tendency apparent in many war-time documentaries away from the stentorian tones of an Emmett towards a warmer, but still authoritative, tone.

Food or Famine

Food or Famine (1962) was directed by Stuart Legg, who had worked with Grierson for the National Film Board of Canada after his apprenticeship at the GPO Film Unit. It was based on an original treatment by Michael Orrom, the Assistant Director on *The World is Rich* and responsible for much of the editing. Filmed in locations across the world, including Europe, South-East Asia, India, South America and Australia, it embodies the internationalism for which the Shell Film Unit was renowned. It is also a good example of the unit's concentration on scientific technique; as one critic remarked, 'the best Shell films are expository psalms to the new technology-ruled world in which the Company stands as so large a symbol' (Robinson 1957a: 10). In style as well as concerns it is the successor to Rotha's FAO film; it is an example of lyrical impressionism which, like many before, uses music in the English style (by Edward Williams) in sympathetic counterpoint to the non-continuous commentary. Shot in colour, it is able to use monochrome sequences to highlight contrasts, especially those with the poor and malnourished. Several voices, of different nationalities, cover a different aspect of the subject, but not within the dialectical interplay of multi-voice commentary.

The introduction establishes the theme; first, there are sequences of many nationalities eating, contrasted with hunger: 'Food is all growth, all vitality. But there has always been starvation … at least half of mankind is malnourished or undernourished.' Picking up the contemporary concern with population growth, the commentary refers to the freedom from want categories: 50 million extra people per annum 'to be housed, clothed, employed, fed'. By the end of the century, it asserts, there will be twice as many people to feed as in 1962. This introduces the film's problematic: 'In a world already underfed, where is the extra food to come from?' But, it asserts, only one tenth of the earth's land area is farmed, one acre per person alive, and not all equally productive.

The second section opens with an FAO meeting in Rome; feeding the world's population is presented as its problematic. It concentrates on actions to render the ten per cent under cultivation more productive; chemical fertilizers, hydroponics, new plant strains, electrically-heated seed beds. An Argentinean voice introduces a sequence on pests; he explains the scientific study of insects to develop new pesticides. A sequence on animal husbandry stresses the necessity for proteins, especially to children. The importance of disease control for animal populations is stressed. A mechanistic metaphor is emphasised: 'Animals are like factories: they need raw materials.' A sequence features the conversion of South Australian desert scrub into grazing by means of analysis of soil and supplementation with trace elements. A cattleman describes it as 'a bloody backroom miracle'. A sequence on ocean fisheries stresses the use of technology: echo-locators, diesel motors, deep-freeze holds, suction-fishing, nylon nets. East Asian fish farming is shown; 'some scientists believe that this farming of the waters, on a bigger scale, is the fishing of the future'. A summary introduces a climax; 'experts believe' that such measures, applied to existing food production, could double it. 'Then why not apply them now, while there is still time?' it asks.

The next section answers this by introducing a composite 'typical farmer of the twentieth century'. 'In his mind there's little to hope for, much to fear. On his shoulders: insecurity, isolation, debt. Into the land he tills, frequently not his, he can put no capital, little knowledge.' The importance of rural education is stressed, using the example of India. It concludes: 'If hunger is to be abolished, it is ultimately the small farmer's will and work that will abolish it. If the extra food is to be grown, this man in his millions must say: you showed me, but I, I grew it [fanfare].'

In answer to a question about whether this will suffice, a section shows the expensive and difficult means to extend the cultivated tenth: in Asia, multi-purpose dams; in Holland, land reclamation; in Venezuela, desalination plants. The conclusion stresses urgency, quoting Orr: 'If we can't agree on this, then there's nothing on God's earth on which we can agree.' The commentary emphasises its technocratic theme: 'Everywhere, the people of the land need more power, water, drainage, fertilizers, plant, machines, buildings, roads, research stations, scientists, advisers, credit.' It concludes with a global citizenship argument that 'we' must provide these things:

> by our work of whatever kind. By the new wealth we create and make available for investment in the soil and seed. We dare not mull over whether this investment is worthwhile. If we fail to make it, as surely as the seasons turn, none will be free from hunger. Shall there be enough food for everyone alive and a reasonable share for each? Or shall the human race return to this? [food riot footage, as *World of Plenty*]. Food – or famine? We have twenty years or less to find the answer.

Conclusion

By 1957 some critics were saying quizzically that 'British documentary of the "interpretative" kind ran suddenly into artistic – as well as economic – doldrums in the period after the war ... it seemed to lack a real sense of direction, a sense of the needs and tone of the times' (Robinson 1957b: 72–3). Scientific filmmaking, however, was thriving in the early 1960s as is clear from the 200 films screened at the Oxford ISFA congress. Documentaries for projection had reached a mature, effective style, and there was little impetus to encourage formal experimentation. In visual terms, the films, frequently in colour – an advantage they had over television until 1967 – often chose international themes where location filming marked a difference from some of the studio-bound television programmes. In aural terms, the experiment of multi-voice commentary was over. Many films continued to have a single narrator or to use several narrators in turn, but increasingly the style of speech typical of current affairs television, with unscripted interviews from participants, was also heard in films.[59] Films (and television too) continued to praise the contribution of science and technology to the modern world, providing answers to problems of human needs; the freedom from want discourse was never far below the surface. Documentary drama was, by the late 1940s, an established style favoured by both the Crown Film Unit (until its closure) and the BBC.

But television was beginning to affect how participants thought about and practised their scientific filmmaking for general audiences. In the first instance, television was perceived as a means of showing complete scientific films; as Bernard Chibnall commented, 'Television can play a very important role in the future in the struggle to educate people in a quickly changing world ... ISFA members cannot afford to ignore the vast audiences within their reach' (Chibnall & Le Harivel 1960: 43). Also, television executives agreed, up to a point; Rotha and Wright's UNESCO film *World Without End* was shown in 1953 for example. Some members of ISFA were concerned with the librarian's-eye-view of the subject; Chibnall, for instance, pondered that 'all producers of science programmes are faced with the eternal problem of wondering where that particular bit of film could be found which would illustrate so well the point they want to make in their programme. There is basically a very real information problem' (ibid.). But television executives had much greater ambitions: whereas science films for projection had stabilised in style and concerns, television had a culture of formal experimentation in the representation of non-fiction subjects and this radically changed the visual representation of science.

6 THE GROWTH OF TELEVISION AND THE REPRESENTATION OF SCIENCE

> Postwar documentarists did not escape from [documentary's] ghetto. The cinema remained off-limits to them and the form only survived in the mainstream of moving image culture because it transferred itself to television. (Winston 2000: 41)

Television both revolutionised the visual representation of science and vastly expanded its audience. Whilst the increase in viewers was the product of broad social changes, the development of the representational styles used for science television was contingent on more local factors, including technical and formal concerns specific to the BBC, and the impact of pressure from scientific groups as they awoke to the public relations power of this new medium. The history of science on television has received little attention (see Gregory & Miller 1998: 41), and this chapter in a book of wider scope cannot rectify this general weakness. The account here is focused on two aspects: the fate of the genres described earlier, and the contexts of the emergence of new, specifically televisual, genres that complemented, and in some cases superseded, some of the earlier forms.

Parallel and interlinked conversations produced the first significant televisual representations of science in Britain from the late 1940s. Competing models of expertise were involved: the programme makers' and the scientists'. On one hand, there were struggles as scientific bodies approached the BBC about who should have the deciding influence over how science should be represented. On the other, television producers were attempting to fashion specifically televisual modes of non-fiction programme making. The result was a new way of representing science on screen that differed from earlier approaches. This approach, created over a space of 15 years, stressed basic science. At first it developed as a contrast with the Rotha model. Then – as discussed in the next chapter – it was forged in the competition between studio and outside broadcast television.

Scientists, science and the BBC, 1941–53

Scientific organisations had shown little interest in the comparatively small scale of scientific filmmaking up to 1950. Both the foundation of the SFA and BA wartime activities signalled an increased concern with science's public relations and its visual representation, but this was virtually at the level of self-help. The relationship of scientific organisations to the BBC, Britain's monopoly broadcaster until 1955, was in a different mode, and similar to the petitioning of government that they had undertaken in 1940 to establish the Scientific Advisory Committee.

Scientists demonstrated a concern, growing over three decades and more, to influence the broadcast representation of science, starting from at least the early 1930s. Where the relations were cordial we may consider this as an analogue, but in a new form, of the networks that sustained the making of documentary science films earlier, including some of the advisory groups that had enabled Rotha's filmmaking. Both scientists and broadcasters had interests in how science should be represented to the public and so it follows that approaches to the BBC by the Royal Society, the British Association and others may be seen as attempts to gain influence for their differing views of science. To begin with, contacts were focused on radio because, before about 1950, television was too marginal an enterprise to elicit much concern.[1] Virtually all coverage of science in the *Listener* at this stage also related to radio broadcasts on the Third Programme. However, both the terms of the debate over scientists' influence on radio, and the principles established under it, came to be applied to television. Not only because of this, but also because television drew on both radio genres and personnel, it is valuable to review these discussions.

The most significant question in the history of relationships between broadcasters and scientists is that of who was petitioning whom at any given stage. It has been normal, across the history of science broadcasting, for programme makers to seek the advice of scientists. In these cases, the final decision on the representation of any programme's scientific content has always been with the programme maker.[2] Agreement was also reached rapidly when the most senior figures in the BBC approached their opposite numbers at the Royal Society about scientific advice in general. But there have also been many occasions on which organised groups of scientists have sought to influence how the sciences and scientists were to be represented on radio and television, attempting to wield authority over the heads of the programme makers. On these occasions, the BBC always retained the upper hand. This account focuses on these larger attempts, particularly because it was in these that the most concentrated debates on the broadcast representation of science occurred. These approaches were contrary attempts at enrolment: of broadcasters seeking to retain the upper hand in control of their medium, and of scientific bodies seeking to control the public relations of science. The currency traded in the discussions in each case was expertise. Harold Perkin's model of professional expertise as a variety of property conferring the security to press a class ideal (see chapter two) may be seen as operating also in competition *between* professional groups, in this case, science and broadcasting. In all these interactions, additional factors were involved, including social and intellectual hierarchy, scientific specialism (whether 'hard' or 'soft' disciplines) and generational effects.

Scientists' proposed mechanisms to control how radio, then television, represented science were, on the evidence of these discussions, not accepted. But there are clear indications that supposedly independent television producers did end up reflecting in their programmes the attitudes of scientists who petitioned the BBC. BBC staff, whilst wary about scientific control, were universally respectful of science and scientists. One explanation for this leakage of attitudes into programmes was the subtle level of influence of scientific advisers to – and participants in – particular programmes; a view of science 'rubbed-off' on producers. This served to amplify the tendency to reproduce views of science circulating in the general culture, of which advisers and programme makers, as much as their audiences, were, obviously, members.

Wartime interactions between the Royal Society, BA, AScW and the BBC set the tone for relations and established the lines of discussion up until the mid-1960s. Over these decades, scientists promoted different combinations of a limited number of mechanisms to enhance – as they saw it – the BBC's scientific coverage. These included scientific advisory panels, appointment of a senior scientist to coordinate broadcasting, appointment of more scientists as producers and, less often, requests to set up a science department. After the Science and World Order Conference of September 1941, Julian Huxley forwarded an AScW memorandum on science and the BBC to Sir Frederick Ogilvie, Reith's successor as BBC Director General.[3] This, taking the line expressed by apostles of scientific planning, stressed the applicability of scientific modes of thought to all types of human problem:

> It is imperative that the general public should be infused with the knowledge that the body of science and its method of development – scientific method – are instruments that can be controlled and utilised to whatever ends a community may choose ... This may be done ... by the rational discussion of problems of social significance which still require solution.[4]

The memorandum suggested a roster of programmes and recommended a scientific committee 'to advise on and develop ideas' for productions. They also wanted the BBC to nominate a Science Programme Officer to liaise between the committee and producers. There had been no scientific advisory committee before the war, but Huxley had attended the Talks Advisory Committee and the physicists Lord Rutherford and William Bragg had both sat on the BBC's General Advisory Committee.[5] The BBC hierarchy ignored the BA's request but, shortly afterwards, Ogilvie wrote to A. V. Hill at the Royal Society asking that, as the Talks and General Advisory Committees were in wartime abeyance, the BBC 'would be grateful if it might be allowed to consult the Secretaries of the Royal Society *ex officio* on matters of this kind. [Staff] would thus have the great advantage of the Society's advice both on plans for broadcasts on scientific subjects and upon the qualifications of particular speakers proposed.'[6] We should note the level of the approach: from Director General of the BBC to one of the most senior figures in the country's principal scientific academy.[7] But the BA were persistent, making a second similar approach after the discussion of radio and cinema at their Science and the Citizen Conference in 1943. They once again proposed a scientists' committee and a science programme officer.[8] The BBC stuck to the existing policy, preferring the prestige of the Royal Society link.[9]

It was not, at the immediately postwar stage, approaches by outside scientists that engendered discussion on the representation of science, but the result of the BBC's concession – whilst under review by the Beveridge Committee (see Briggs 1979: chapter 4) – that they should reconfigure the membership of the General Advisory Council. This was reformulated in autumn 1947 as part of wider changes after Sir William Haley, then Director General, had been subject to criticism from the Prime Minister, Clement Attlee, about political reporting. Haley ensured that the new Council covered 'a very wide variety of representative interests – Commonwealth affairs, the Press, science and humanities' (Briggs 1979: 126–7). Seven of the 49 members had scientific qualifications or experience. One of these, Mark Oliphant, Professor of Physics at Birmingham University, initiated a discussion on the broadcasting of science in 1949. As a particle physicist, he had been involved in electromagnetic separation of uranium isotopes at the University of California, Berkeley, alongside the Manhattan Project, and had subsequently become a vocal opponent of nuclear proliferation (see Bleaney 2006). This interaction between scientists and the BBC – which stretched over three years – explicitly voiced the terms of the debate on the representation of science as it evolved for over a decade afterwards. It also marked the first point at which television was explicitly in the frame in terms of approaches by scientists to affect the representation of science in BBC broadcasts. In his intervention Oliphant was a standard-bearer for what he termed fundamental, as against applied, science. He was anxious that science should not be made the scapegoat for problems caused by the bomb, and suggested the appointment of an advisory committee on scientific broadcasting and the strengthening of scientific programme production staff.[10] Oliphant was quite clear on how he felt the balance of programmes should shift away from the social:

> I would like to see ... some break away from the perpetual theme of 'science and society', with the inevitable excursion of the scientist into fields of politics where he does not shine, towards an attempt to present science as natural philosophy, as a way of life and a culture in its own right ... I don't think scientists should always appear as Utopian idealists, as Marxists, or as amateur politicians. Cannot we sometimes forget war and atomic weapons, industrial advance or productivity, medicine and food production or science and religion and say something more of the history and growth of science, of the great revolution wrought by the introduction of the experimental method, of the intellectual satisfaction and fun of science and of the scope and content of modern science.[11]

We may note that this initiative did not start with demands for a scientific adviser, but with opinions about how science should be represented to the public. The contrast with the established conventions of representing science as socially useful, as seen in the films of Paul Rotha, or as represented by the wartime delegations from the BA and AScW, was significant. Oliphant's criticism is congruent with Gary Werskey's conclusion that 'Bernalist' or radical conceptions of science were eclipsed in the postwar period and replaced with concerns about the freedom to pursue 'pure' science (see 1988: 281–5). Broadcasting, in Oliphant's view, did have an urgent responsibility in relation to science, as he argued at a meeting of the BBC's General Advisory Council:

> The evil wrought by science springs, not from any intrinsic evil in science itself, but from its misuse by men who do not really understand what science is. It must, therefore, be one of the primary aims of the BBC to find a medium for the rapid education of the public towards a properly-balanced view of what science is, how it works and how it affects the lives of all men, while at the same time emphasising its limitations and what it might achieve in the future.[12]

This approach, from a 'semi-insider', compelled the BBC into more action than its previous covert but outright resistance to scientific control. A subcommittee, chaired by the distinguished civil servant Sir John Anderson, was appointed to discuss the issues that Oliphant had raised. Members were

Oliphant, A. V. Hill, the civil servant Sir Alan Barlow (who sat on the government's Advisory Council on Scientific Policy) and the physicist and crystallographer Sir Lawrence Bragg, then running the Cavendish Laboratory in Cambridge. They believed that that there was a 'considerable potential audience for broadcasts on the lines proposed by Professor Oliphant':

> Broadcasts should be simple at the outset. They should be aimed at an audience consisting of people with very little scientific knowledge but with an interest and also people who already have some knowledge of the subject and who might like to hear more about it from an authoritative speaker. It would not be part of the aim to attract listeners whose interests lay entirely away from all such matters.[13]

It was in this period, when radio – especially the Third Programme – was broadcasting quite extensively on scientific topics, that a new factor, that of comprehensibility, began to be perceived as an issue. The BBC's Further Education Department had conducted an investigation into how much its intended audience – the general public – could follow of a radio programme, 'What We Do With Electrons' (21 July 1949). The speaker's 'talk was imperfectly understood, or not understood at all, save by a limited group of listeners with a good background of scientific knowledge'. It conceded, however, that '*interest* was not confined to any one group but was found in all and was, in fact, most strong where the content of the broadcast was only partly understood' (emphasis in original).[14]

The sub-committee,[15] very struck by this research, devoted more than a third of their report to the questions of audience that it provoked, discussing a scale of comprehension mapped – in declining size of audience – to the three radio stations, the Light Programme, Home Service and Third Programme.

Their solution to their perceived problems of broadcasting was, once again, for the BBC to appoint a scientific adviser 'of high standing' attached to the staff of the BBC to represent all output departments and to ensure the establishment of 'high-level contact with scientists on the requisite footing of informality'. The brief would include: coordination of science broadcasts in the various programme services; establishment of personal relations with scientists and scientific institutions to get the best available speakers and presentation of suitable topics; exploration of programme ideas and study of method and technique of presentation; study of listener reaction and exploration of potentialities of television programmes in this field.[16]

The BBC accepted the Committee's report and set about selecting a suitable candidate for the adviser role. Jacob Bronowski and C. P. Snow were considered. The idea of appointing Julian Huxley was briefly entertained, but rejected by both broadcasters and scientists. George Barnes considered him to be 'an obvious name', but he judged that 'he has probably been out of science for too long too recently [sic] to feel that he was sufficiently in touch with the scientific world'.[17] A. V. Hill agreed: 'I wouldn't support J. H., filled with bright ideas but unreliable.'[18] Barnes suggested that 'Sir Henry Dale, if he would consider such a post, would be an obvious choice as he has given us a great deal of advice both on the talks and the features side in the past'.

Accordingly Dale, Nobel Prize-winner, distinguished biologist, Director of the Wellcome Trust and then aged 75, was appointed from July 1950. If the BBC's views were dominated by radio, this was still more the case with Dale. As far as his notion of the capabilities of television was concerned, he was true to the accepted view that 'the major responsibility of Talks was to present eminent speakers on chosen subjects' (see Scannell 1979: 98). For Dale, television science was like a popular science lecture but with a larger audience:

> A lecturer expounding science directly to a mixed audience would seldom attempt to do this by word of mouth alone; habitually he would employ visual aids to description and understanding – blackboard, wall diagrams, lantern slides, films, or the actual performance of experiments. All these, or their equivalents, can be transmitted by television. (1950: 140)

Dale served for two years as adviser, with rather inconclusive results. He generally approved of the little science that did appear on television, perhaps because, as he reported, he 'was, on the whole, able to cooperate more fully in this enterprise than that comprised in the various sound-broadcasting problems, which, being longer established, tend, naturally, to adhere more tenaciously to methods already traditional'.[19] He concluded:

> It has seemed to me that television has a great opportunity, if it starts with such basic facts of experimental science as can be illustrated by simple but striking demonstrations, and builds up from this foundation to a working conception of the more complicated uses of scientific knowledge, in the equipment of ordinary life in this scientific age – including, in particular, a simplified understanding of the scientific devices involved in television itself.[20]

Here he echoed the audience concerns of the Anderson sub-committee. Two of Dale's significant and specific comments related to different aspects of the balance in programming: between varying subjects and over the appropriateness of a stress on current science. He commented that 'in general, I am inclined to the view that they tended to aim at too much elaboration of particular themes, with a preference for medical and psychological subjects'. This shows that participants' definitions of science included medicine, but that, even at this early stage in the history of television, questions of balance pitched the physical sciences, medicine and applied science and engineering against each other.[21] The second issue of balance related to that between a type of basic scientific literacy and science news. Dale's stress was more on scientific modes of thought than on television reporting of science in action:

> A sound education of the public, in an understanding of the scientific revolution in which we are living, requires, in my opinion, a much greater emphasis on the historical origins and the principal stages in the development of modern science, than broadcasting programmes have hitherto been able to give. I believe that *lectures and films*, dealing with these chief events in the history of science ... could be most attractive (emphasis added).[22]

He echoed here Oliphant's original stresses on the history of science and the experimental method. But, in relation to genre, he was out of tune with the ethos of producers.

This was the end of the BBC's rather grudging experiment in the employment of a scientific adviser. Mary Somerville, Controller of radio talks, who had worked closely with him, described the circumstances of his appointment and their treatment of him as 'a somewhat unhappy page in BBC history'. She added that 'although the BBC accepted the recommendation of the Anderson Committee to appoint a Scientific Adviser I do not think that they ever subscribed to the assumptions underlying that recommendation i.e. that it is desirable in principle to co-ordinate BBC output, or approaches to scientists, or that it is necessary for any check to be placed upon producers' choice of speakers by "establishing high level contact with scientists on the requisite footing of informality"'.[23] She added: 'Even if we did decide to have another standing scientific adviser on the former basis, I would not recommend the re-appointment of Sir Henry Dale. He is too old for the job, and in our present circumstances there is really no need for a scientific adviser.' Even gerontocracies have their day, it seems. Somerville was of the view that Archibald Clow, a scientifically-qualified radio producer could serve perfectly well in the role of adviser.[24] Dale may have left the BBC, but his model of science broadcasting focusing on basic scientific literacy, congruent with Mark Oliphant's earlier view, was gaining ground in some of the actual programmes being made.

Televisual means and genres: the invention of science television

Television programmes received only slight attention in the formal interactions between scientists and the BBC up to Dale's report in 1953. But the 1950s was the decade in which the medium began to seem – to programme makers, filmmakers and scientists – the natural medium for scientific programming, as numbers of viewers increased dramatically. Television was a medium finding its feet in the two decades after broadcasting recommenced in summer 1946, as it was converted from a luxury service for the few, broadcasting only a few hours per day, to a popular medium with millions of viewers. There was a more than 300-fold increase in television licence holders between 1947 and 1955, by which date 4.5 million homes had sets (see Briggs 1979: 240). Science on television developed as a special case of non-fiction television in general. As the medium and genres within it developed, technological capacities interacted with institutional factors with the BBC's staff working to produce new types of programmes.

Technological factors, especially the distinction between film and live television, played a significant role. In the 1950s, television was mainly a simultaneous medium – virtually all the output that viewers saw was being performed at that moment, usually in the television studios at Alexandra Palace in north London. 'Outside broadcasts' were, in a sense, only the epitome of this simultaneous medium, although highly significant in the scope they gave to include actuality in television output. The only exception to this simultaneity was when a film was 'telecined' – broadcast via a specialised television camera that scans each frame of film. There was no way in this period to record live television other than, from 1947, 'telerecording' – the reverse of telecine, where a film camera shoots the image on a television monitor – which gave rather poor results (see Briggs 1979: 278). Videotape recording was introduced from 1958 but did not make a significant impact on televisual forms for some years.[25] The large size of cameras also dictated that, for most of the time, it was most convenient to shoot programmes in the studio. It is a conventional trope in writings about documentary that the introduction of lightweight equipment enables new freedom in production; this was said of smaller film cameras in the 1930s and it was said again of the introduction of smaller television cameras in the late 1950s (see Bell 1986: 80).

This is not crude technological determinism; contemporaries actively felt the constraints. Cecil McGivern, Controller of Television Programmes from 1947 to 1957 (see Briggs 1979: 223–4; 1995: 15),[26] said that 'at present, much of [outside broadcast's] equipment, though good, is still too ponderous. Extra lightweight cameras are needed, each camera containing its own transmitting equipment' (1950: 149). He also had an acute sense of how the types of television sets most widely used in the home had an impact on the styles of programme making. With 62 per cent of viewers watching on sets equipped with only a nine-inch screen,

> Good television with screens as they are today means close-up ... the television producer ... knows that the human face and body in long shot on 9-inch screens mean nothing at all ... So he manoeuvres his cast, he steers his cameras, he twists the action and arranges his groupings, he gets up to all sorts of dodges and wheezes ... in order to get his actors in to close-up ... He eliminates passages from scripts that prevent his doing this ... the Television Service itself eliminates good ideas, good scripts, good programmes which would prevent the producer doing this. (1950: 146–7)

Such technical factors affected discussions about the form that television genres, including non-fiction programmes, should take. Programmes could be made as one-off performances, either outside broadcasts or based in the studio with elaborate sets, presenters, interviews and filmed inserts. The alternative was to make films that were complete before broadcast via telecine and able to be repeated.[27] As the producer Mary Adams argued in May 1946, a month before television restarted:

> It is important to distinguish between film and direct television in considering how television can best be used in the future. In many ways, film has an advantage over television; material can be cut and arranged, and time and distance overcome. The television cameras, however, put a greater emphasis on the skill and personality of the demonstrator, and produce a heightened sense of actuality and immediacy. But the telescoping of lengthy experiments in short sequences is a tricky business for television.[28]

In the longer term, of all these elements, it was the personality issue – linked to factors of immediacy and 'intimacy' – that became significant in non-fiction broadcasting.

For nearly two decades after the war, the BBC mainly organised its approach via categories of medium rather than content; science was one of many subjects that might be represented by the several different departments that made non-fiction programmes. There had been some coverage of science on television before the service was closed down for the duration of the war: experiments were demonstrated in the studios; drawings, diagrams, pictures and scientific films were shown; episodes in the history of science were dramatised and scientists appeared. In sum, 'many experiments were made in the presentation of scientific material, although no coherent programme of educational telecasts was undertaken'.[29] But this 'infant service' gave scant impression of how it would develop after the war. The institutional position of television documentary immediately postwar has been described as 'a small part of an isolated and under-funded service regarded with at best indifference and at worse suspicion and hostility by the senior members of a broadcasting monopoly firmly rooted in the traditions of sound broadcasting' (Bell 1986: 70). During the war, Sir Allan Powell, chairman of the BBC's Board of Governors, betrayed the way that television was considered the junior partner when he commented that 'it seems obvious that television must be the outstanding direction in which the *art of radio* will progress' (emphasis added).[30]

Television was effectively a new medium in the postwar period, and those involved experienced the pressures of establishing it; McGivern said that 'just to keep it going is a headache ... Television is young, but it is already swallowing at a frightening rate the output of writers, producers, designers. How are we going to keep it fed?' (1950: 142). His proposed answer was to limit the hours of broadcasting (see 1950: 144). It is notable that, for him, actuality provided the means to escape the feeding frenzy of studio-based television: 'There is, however, one category of television programmes in which the task of finding satisfying material is – and will continue to be – much easier. That is the outside broadcast, the very essence of television ... Go where you may, there is a subject for the television outside broadcast camera' (1950: 148–9). He added that the film camera, too, was crucial to providing output: 'Though not possessing the immediate, vivid interest of the outside broadcast camera, it can penetrate when and where the other camera cannot'; here he was looking for distinctive use: 'Like the Outside Broadcasts Section, it calls for constant development. Its equipment and methods should be devised so that they suit the speeds and finance of television' (1950: 149).

It was only when television gained, in October 1950, full departmental status, a Controller and its own seat on the Board of Management, that it began to overcome its rather lowly status relative to radio (see Bell 1986: 68–9). However, the 1947–48 reorganisation that coincided with the changes to the General Advisory Council discussed above had also begun the differentiation of the departments that latterly produced several genres of television science programme. In December 1947 the television workforce, previously in one large division, began to be divided into different departments covering drama, light entertainment, talks and talks features, and 'documentary and magazines' (BBC staff lists; see Briggs 1979: 227–8). Under the BBC administrative structure that divided all these departments producing programmes (Supply) from their planning, commissioning and funding (Output), each was in competition with the others for broadcast slots (see Bell 1986: 67–8).

The different representational styles and traditions that these departments stood for set the terms in which the BBC organised its approach to television in general, to non-fiction, and to science programmes. Within science, technology and medicine the Talks Department had greater coverage of the physical sciences, applied science and engineering, whilst the Documentary department showed a preference for medical subjects. But the gravitation of these different technical subjects towards the separate departments did not reflect a deliberate decision based on subject matter. Rather, it must be concluded that it was the product of deeper assumptions about the respective character of science and medicine; one as predicate knowledge suited to lecture-like exposition and the other as intrinsically social. Talks and documentary, in their prewar forms, had already been aligned on the two sides of these assumptions.

The BBC's Talks culture, derived from that of radio, was the foundation for non-fiction television. Accordingly, the Television Talks Department was the main centre for scientific programmes for most of the 1950s, specialising in studio-based live television shows. (The Outside Broadcast (OB) department also produced a small number of programmes representing science, technology and medicine in this period, where the coverage was more closely aligned with Talks than Documentary.) Robert Barr, a documentary producer, conveyed the spirit: 'In a TV "talk" expert opinion or information is conveyed directly from the authority to the viewer. It is Haldane talking on science, Harbin demonstrating his methods of cooking ... public men personally discussing events in the news.'[31] On the evidence of the Anderson committee episode, science and engineering would seem 'natural' topics for talks. Broadcasters were, unsurprisingly, acceding to wider assumptions within society about the abstruseness of science, beliefs that were shared by many laypeople and scientists alike.

By contrast, the BBC and its staff, in establishing a documentary department for television, retained the prewar association between documentary forms and social interpretation. It is perhaps not surprising that medicine – often with an emphasis on public health – was favoured for representation within the output of this department, as it was seen to provide the human-interest stories and social issues associated with the form. Accordingly, when Rotha became Head of Documentaries, the department only had particular social aspects of science, technology and medicine within its remit.

Television Talks: studio-based science programming

In the Television Talks Department, the pressure to innovate in televisual terms operated in a context in which models that originated in radio were translated into new visual forms. The department developed a number of formats for science and technology programmes in the period discussed here. Its managers and producers came from a diversity of backgrounds. Mary Adams, its head, was a Cambridge scientist taken on by BBC radio after giving a series of talks in 1928; she had become one of only four television producers before the war (see Miall 1994: 72). The producer Andrew Miller-Jones had directed films for Gaumont-British Instructional prior to joining BBC television before the war. A significant producer of science programmes was James McCloy, who joined the BBC schools service in 1949, described himself as 'a trained scientist with research experience' and saw himself as a specialist in the presentation of science to the public.[32] In fact, he had a first degree in zoology and had not worked in scientific research (see Grattan 2002). George Noordhof had been a research scientist before joining the BBC (see Forbes 1954: 10).

Inventors' Club

The series *Inventors' Club* exemplifies one end of the broadcasting spectrum, namely presenting technology as light entertainment. In this series, shown between April 1948 and December 1956,[33] 'a couple of baggy-suited handymen displayed working, practical, sometimes highly commercial inventions sent in by viewers' (Black 1972: 20).[34] First produced by Andrew Miller-Jones and later by James McCloy, it was broadcast live from the studios at Alexandra Palace with an occasional insert film of inventions in action. It owed something to the quiz panel discussions of *The Brains Trust* (1942–49) or *Animal, Vegetable, Mineral* (1952–59), but with the addition of lay participants. Here, inventors demonstrated their creations with the assistance of the presenters Leslie Hardern (of the Gas Light & Coke Company and a broadcaster who had specialised in domestic design) and Geoffrey Boumphrey ('who had been in turn engineer, cook, housing expert, town planning expert and practical inventor' (Hardern 1954: 25)).[35] The series was considered by the Talks Department to be broadly 'scientific', as is indicated specifically by Mary Adams' wish to have Professor Andrade 'or another authoritative scientist' and a member of the Parliamentary Science Group on the panel.[36] The category 'inventor' denotes a very particular type of technologist (see chapter one). By this period, an inventor was generally a layperson, although the programme deliberately set out not to confine itself to amateurs, but to include also pre-production items invented by those working for organisations.[37] The type of ingenuity displayed in the cohort of 7,000 applicants – of whom 500 appeared in the shows in the first six years – included domestic gadgets created in the spirit of the wartime campaign

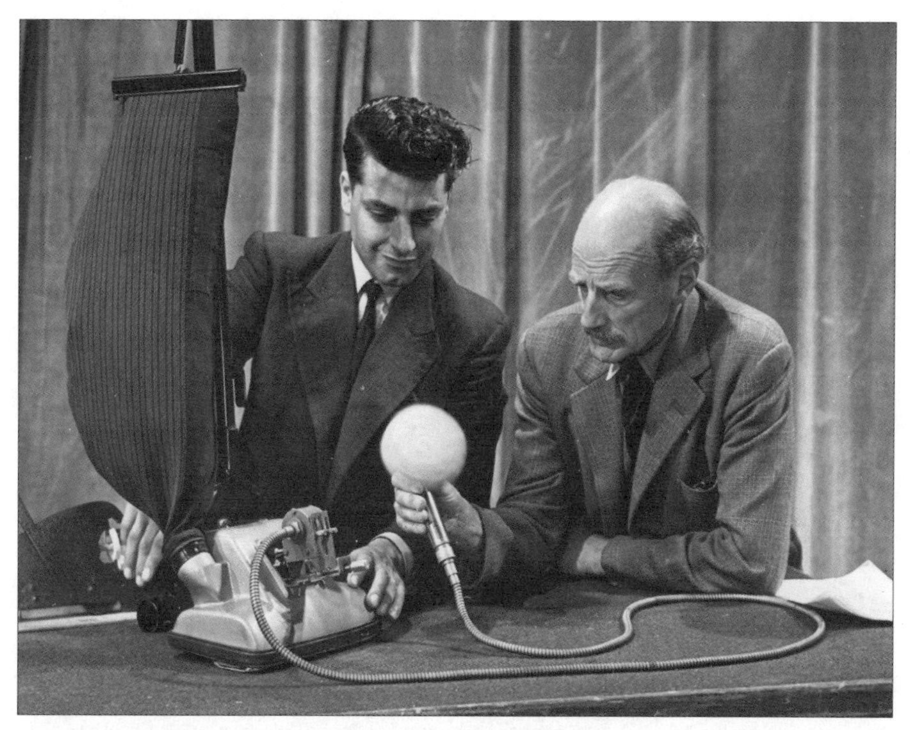

Fig 6.1 ***Inventors' Club*, 1953: presenter Geoffrey Boumphrey (r) with Mr Richards of Torquay demonstrating his two-way gear drive for vacuum cleaners**

to 'make do and mend' and suggestion-box styles of innovation to workplace devices and practices. Early programmes included, for example, rubber table-corner buffers to save the heads of toddlers (suggested by a newly-demobbed soldier) and a device invented by aircraft draughtsmen to produce isometric drawings: 'using pieces of Meccano and of bicycles and wire and string, they had built a machine to do it for them' (Hardern 1954: 24). The show also had its share of perpetual motion inventors. Amateur, independent, practical and innovating by increment and combination of existing mechanical principles and devices; this was the world of the inventor revealed in these popular programmes. Starting in the years when Stafford Cripps was seeking to enhance British economic performance and efficiency (see chapter five), here was science television representing lay engagement with technological improvement presented as popular entertainment (see Hardern 1954: 3). This had been the intention from the beginning, as an internal memorandum commented that 'the object of the programme would be to stimulate and promote inventions and good designs likely to assist exports and encourage the home market'.[38] By 1953, ninety inventions featured in the programme had gone into large-scale production.[39] It achieved some recognition of this kind when it was cited in a House of Commons debate on productivity in 1956.[40] The enterprise does not need to have been entirely serious-minded to function as a public representation of this particular mode of science and technology. Hardern also expressed an awareness of the line he was walking in representing invention:

> If the office staff were expecting the kind of eccentric gentleman one so often sees in cartoons depicting inventors, they must have been disillusioned. Personally speaking, our inventors were smart and businesslike, although our commissionaire might be forgiven for an occasional furtive or apprehensive eye cast at their intriguingly shaped parcels. (1954: 25)

In fact, the appeal to the audience was based on balancing four factors: the excitement of the unexpected; 'the human appeal of the ordinary man or woman appearing on the screen'; 'the powerful sympathy aroused for the "little man" striving to win support'; and 'the intrinsic interest of the inventions themselves'.[41]

The live aspect of the show, with its panel of experts judging the efforts of lay people, coupled with the inventions' capacity to fail or to cause unscripted problems (as when chickens enjoying a new design of feeding trough drowned out the sound of the presenters) all contributed to this programme being at the lightest end of the presentation of science and technology. Recalling the categories of the Anderson sub-committee – which met at the beginning of the programme's eight-year run – this was a show which, had it been a radio broadcast, would have been on the Light Programme, attracting a large audience, but with only the slightest of scientific content.

Unlike *Inventors' Club*, most of the Talks Department science output was limited to single programmes or series of one run. A few examples will give an idea of the range. Andrew Miller-Jones produced two studio-based programmes on scientific films in 1948 and 1953, which drew on the congresses of the ISFA, showing such examples as Jean Painlevé's *Formation de Cristaux* (1950) and Margaret Thompson's *Prevention of Cross Infection* (1952). Filmmakers, including Painlevé and Arthur Elton, introduced excerpts that were played in by telecine.[42] Miller-Jones produced 13 programmes in the medical *Matters of Life and Death* series between 1949 and 1951, with subjects including atomic medicine, deafness and tuberculosis. Reproducing in new form the excitement of the Urban-Duncan films, television cameras were attached to a microscope for some of these programmes, which also used scientific films.[43] *Science in the Making*, which ran in 1953–54, produced by George Noordhof and presented by Jacob Bronowski, was an early example of audience participation. Viewers were encouraged to send in data; for example, for a programme on twins the *Radio Times* published a set of questions to which twins were invited to respond, about upbringing and susceptibility to hay fever and asthma (Anon. 1953b: 15). Other programmes discussed the common cold, left- and right-handedness, and 'the living memory', featuring the neurologist William Grey Walter.[44]

Science outside broadcasts

The Outside Broadcast Department prewar had specialised in televising events that were happening already, such as state occasions or sporting fixtures. In this way, it was distinguished from studio-based television, which was more constructed. Formally separated from the film facilities department in 1948, it was involved in the production of occasional science programmes. One of the plainer examples was a broadcast from the Science Museum's new aeronautics gallery in September 1950, in a series of museum visit programmes. Produced by Alan Chilvers from the OB department, it was comprised of Raymond Baxter (an ex-Spitfire pilot) asking the museum's keeper of aeronautics simple questions such as 'how did flying begin?', to which W. T. O'Dea gave lengthy answers illustrated by shots of the exhibits.[45] More complex were the two broadcasts from the Pioneer Health Centre, Peckham, in 1950. In these cases, research and production was shared between Duncan Ross from Documentary and Chilvers from OB.[46] Other OBs came from British Association meetings, for example in 1950 and 1954.[47]

Documentary Department science and medical programming

The Documentary and Magazines unit was another of the departments formed at the end of 1947. Documentary was perceived by the Controller of Television, Norman Collins, to be an entirely new genre for television, starting from 1947, requiring substantial extra resources of multiple sets, actors and film inserts, over and above the simpler requirements of talks. But he maintained that there were 'many subjects ideally suited to documentary treatment – programmes on medical subjects, programmes on legal procedures and programmes of an historical character … – which are incapable of being satisfactorily treated in anything other than documentary fashion'.[48] The department's staff came from a variety of backgrounds. They included the producers Robert Barr (from journalism and radio) and Stephen McCormack (from theatre), along with the scriptwriter Duncan Ross (from working on Rotha's newsreels), and various 'guest producers' (see Bell 1986: 71; Briggs 1979: 282; BBC staff lists). Norman Swallow, an erstwhile journalist who had worked in the radio features department at BBC Manchester,[49] transferred to the department in May 1954 from Talks – bringing his current affairs series *Special Enquiry* with him. Cecil McGivern, Controller of Television Programmes, expected the department to have a major and distinctive role in establishing kinds of non-fiction television that went beyond the Talks formats. In April 1951, for the first time, a Head of Television Documentaries post was created but, despite several attempts, it was not filled for two years. In a letter responding to an expression of interest in the job from Paul Rotha, McGivern described it as 'the most important and most fascinating job in Television'. He said that 'an important part of the development of television lies in this man's hands'. At stake was the definition and potential of documentary within the medium: 'in my firm opinion the development of television documentary does not lie *only* in the development of filmed television. We have much more to do at present than simply aim at perfectionism' (emphasis in original).[50] Here he was referring to the development of new, specifically televisual, forms in distinction from using television as a means to transmit specially-made films. McGivern was also, however, committed to the use of films, as he wrote to George Barnes (Director of Television Broadcasting) in August 1952: 'There are certain subjects which can only be adequately represented by their being all on celluloid ... I am sure that ... any Head of Documentaries would not be satisfied with his output and coverage until he had achieved complete programmes on film.'[51] But he saw them as an expensive option, needing to be scheduled sparingly.

The self-awareness of staff in defining their medium derived from the fact that it was seen – despite some minor experiments before 1939 – to be a novel feature of postwar television. Producers felt an obligation to distinguish it from Talks, the dominant existing non-fiction genre. Senior staff placed a premium on form in the infant television service, and the documentary department was expected to create novel ways of representing actuality. Members of the department wrote frequent memoranda and articles on documentary's distinguishing features. One of them acknowledged that 'the conflict of ideologies in television is a regular reminder of its infancy and is like children with a three-legged stool arguing as to which leg is the most important' (Ross 1950: 20). Their differing understandings reflected not only their backgrounds, but the absence at this stage of firmly established formats particular to television. They agreed that the role of their competing department, Talks, was to present eminent speakers:

> On television these talks are illustrated because it is usually more interesting to see what a person is talking about rather than merely to see him talking. Whatever form of illustration is used ... the first duty of 'talks'

is, obviously, to interpret a speaker. Documentary, on the other hand, is not concerned with speakers and is at its best when it ignores them completely and interprets the subject straight from life. (Ibid.)

This explicitly-defined documentary is in distinction to the way that scientists such as Henry Dale expected television to show science; whatever science documentary might become, it would not be its role merely to convey scientific lectures.

Duncan Ross, the film documentarist, placed television documentary firmly within the Griersonian 'creative treatment of reality'.[52] Reflecting on the more and less successful television documentary genres as early as March 1948, he had perceived four types of programme. Industrial documentaries were inappropriate because industrial machinery could not be brought to the Alexandra Palace studio; in any case, 'recent COI and commercially sponsored films have tired the public mind with dull repetitious visuals of factories, forges and foundries'. The ease with which he made this judgement, only 15 years after Flaherty's *Industrial Britain*, is a measure of how prewar concerns were being rejected by new generations of film and programme makers. 'Progress Report'-style programmes, he continued, were likely to require too large a staff to be economical, especially given that the ratio of subjects worth televising after research was likely to be rather low. The two genres he favoured embodied the split between films made for transmission and studio-based live television; he described them as soundly researched documentary and story-type documentary.[53] Of his last type, known in the department as documentary-drama, he said 'with only a few facts they can be written from the imagination and trimmed to fit the studio floor'. But, he added, they 'are less informative than "straight" documentary, often tend to become mere dramatic fiction, and good subject matter for them is limited'.[54]

Robert Barr, by contrast, with his newspaper and radio background, weighted his definition towards drama for, as he said, documentary

> takes into account many opinions; its nature is to select, edit, synthesise and present its own conclusion ... [It] is concerned with actions; its form is the dramatisation of facts, reconstruction of events, and it uses any dramatic device to make its point. It will use (and devise) any technique that will give force and clarity to the information it seeks to convey. Its intention is to make people feel as well as think. Its appeal is to the emotions and it talks in terms of human conduct.[55]

In his view, 'documentary gets its dramatic effects out of truth, out of the audience accepting it as being true'.[56]

Modes of documentary programme making

The Documentary Department produced programmes on a wide range of subjects: 'medicine, delinquency, handcrafts, police, theatricals, crime, courts, politics, mining, textiles, science, education, social experiments and many other subjects' (Ross 1950: 21). It was also responsible for several different genres of non-fiction, including films on artists, documentary-drama and 'documentary magazine' programmes. The latter category was formed of the series *London Town* (1950–51), *About Britain*, *Special Enquiry* and *The World is Ours*.[57] The former included several drama-documentaries on medical and health subjects that would have fallen within Rotha's wartime definitions of scientific films: *To Save a Life* (written by Rex North and produced by Robert Barr, 25 June 1951), *Dangerous Drugs* (written by Cormac Swan, produced by Barr, 11 April 1952), *Family Doctor* (written by Barr and Swan, produced by Barr, 3 September 1952), *Under Her Skilled Hand* (on the teaching hospitals and nursing, written by Swan, produced by Barr, 4 February 1953), *Outbreak* (about foot and mouth disease, written by John Irving and produced by Barr, 3 March 1953), *Behind these Doors: Doctor's Waiting Room* (written and produced by Duncan Ross, directed by Gilchrist Calder, 20 January 1954) and *Medical Officer of Health* (produced by Barr, 21 September 1954).

The example of *Medical Officer of Health*[58] shows how the documentary drama model worked. The aim was to construct an ideal-typical story that represented the subject by creating fictitious characters to perform a narrative that conveyed the chosen social themes. The technique had become well-established in postwar documentary (with, for example, *The Centre*) and producers were aware of the precedent (see Doncaster 1956: 44). Barr scripted the programme by building it around the dramatic experience of 'Dr Smith', a fictitious Medical Officer of Health (MOH) dealing with a smallpox outbreak in an industrial town, 'Weirsley', an ideal-typical construct conveyed by filming inserts in several different towns. The research started with Robert Barr reading a journal account of a real outbreak in Todmorden. He went on to discuss the outbreak with its two authors before constructing his narrative.[59] Barr's report explained how he aimed to convey the work of a typical MOH by using the dramatic structure provided by the outbreak. Viewers would see how Dr Smith and colleagues traced and vaccinated the contacts of the mill worker who had contracted the disease. The lack of financial provision for contacts to stay off work would be conveyed. The rationale of the policy against general immunisation of the population could be explained whilst the drama covered the development of the outbreak to 26 people in six weeks. The – ultimately unsuccessful – attempts to trace the source of the original infection were also to be covered.[60] Like virtually all such documentaries, *Medical Officer of Health* was performed once for transmission by actors who received no billing in the *Radio Times*, a policy that was intended to reinforce the illusion of actuality. In this, the makers of these documentaries were convinced that these fictionalisations were realistic, even 'true' (see Bell 1986: 74–7). The audience appreciation index (effectively a percentage quality rating by viewers) showed a very high level of eighty.[61]

The main vehicle for social reportage current affairs was *Special Enquiry*, which ran from October 1952 and every autumn until 1957, produced by Norman Swallow (see Swallow 1966: 72). Introduced by the Manchester journalist Robert Reid, *Special Enquiry* was an early example of 'current affairs', with the structure mirroring the social reportage of print journalism. It had the innovation of a journalist – in this case deliberately chosen for his regional accent – acting as the viewer's proxy; the programmes had the implication that the viewer accompanies the journalist as he makes his investigation on their behalf (see Corner 1991b: 44–5). Amongst its subjects were housing issues, social medicine and smoke pollution (all of which had previously been the subject of documentary films).[62] As Swallow stated, picking up the genealogy with the 1930s gas-sponsored documentaries: 'They are all social problems' (1956: 54). The programme on coal smoke, broadcast in the first series in January 1953, was typical in format. The connection with John Taylor's film *The Smoke Menace* (1937) was clear, not simply in the re-use of the title but also in the explicit showing of an excerpt in the introductory section of the programme.[63]

Rotha at the BBC

For two years from May 1953, Rotha was Head of this Documentary Department. BBC executives were so eager to employ him that he successfully drove a hard bargain on his salary.[64] He replaced the documentary-drama specialist Robert Barr, who had been acting Head of Department for six months, an unsuccessful candidate for the post. The department was renamed 'Documentary Programmes' around the same time and several more staff were added, including Swallow.[65] Rotha entered into the job with some enthusiasm, as he had come to appreciate that a television programme could achieve many more viewers in a single broadcast than a film would manage in months of distribution: 'Ephemeral as a one-night stand television may be, but to counteract its sudden birth-and-death, is its fantastic simultaneous access to a mass audience under conditions wholly different from the movie in the cinema.'[66] In this period, the department was producing about 35 programmes a year of an average length of 45 minutes (see Rotha 1955: 371).[67] He was reported in *The Star* as saying 'TV appeals to me because it gives such a wide window on the world to the working man as well as the middle and upper classes. I want to get great breaths of fresh air coming through that window'.[68]

It is clear that when Rotha came into the department, he shifted its emphasis away from Barr's documentary-drama mode – which was in the ascendant – towards documentary films and current affairs. If we accept their categories, this shift was significant. The producer Caryl Doncaster argued that 'the talk informs. The dramatised story documentary interprets. One appeals to the intellect, the other to the emotions' (1956: 44). In Rotha's prewar definitions, impressionistic documentaries appealed to the intellect via the emotions. But, given that the documentary-drama was the main genre of the unit up to this point, the change postwar signalled a move towards a greater intellectual than emotional appeal. One of the department's distinctive products during his reign was the 12-part series, *The World is Ours.* Although not specifically identified as a scientific strand, many of its subjects were of a piece with Rotha's older model of documentary films showing science as the means to achieve freedom from want. After the first three issues made, somewhat like *Special Enquiry*, in the studio with elements of documentary-drama and inserted film sequences, the remaining nine programmes were produced as complete films and telecined for broadcast. The series was made in conjunction with UNESCO and represented the work of the United Nations agencies, reproducing the internationalism of postwar documentary and the social relations of science agenda. Programmes included: *World Nurse* (on World Health Organisation (WHO) nursing services, broadcast April 1954), *World Population* (with emphasis on food supply, June 1954), *The Wealth of the Waters* (exploitation of fish supplies, August 1954), *The Invisible Enemy* (on the WHO's virology programme, May 1955) and *Wealth from the Wilderness* (on the greening of deserts, August 1955).[69]

Despite Norman Swallow's series producer credit there is no mistaking the Rotha signature on these films, although they did not use his distinctive multi-voice commentary, still less dialectical montage. In subject they were explicitly in the mould of his previous films, *World of Plenty, The World is Rich* and also *Land of Promise.* This is particularly clear in the *World Population* programme, which examined the relationship between population increase and food supply, featured the FAO, and conveyed a series of technical means to improve matters. We might think of it as sitting between *The World is Rich* and *Food Or Famine.* Where these programmes do not look like the longer Rotha films, they certainly bear a close resemblance to his newsreels. This stylistic similarity extends to the use of Rotha's own footage in the title sequence, some of it shot twenty years before. Rotha also employed a significant number of old associates to work on the films, including Peter Ritchie

Calder to write scripts (notably for the world population programme), along with Michael Orrom, the assistant director from Rotha's company (who scripted the fourth programme, about refugees), and Basil Wright (co-writer of *World Nurse*), the documentary filmmaker and co-director of their joint film *World Without End* (1953). He also employed Marie Neurath to produce Isotype diagrams.

The Invisible Enemy, produced by Swallow and scripted by documentary veteran Jack Chambers, can be taken to exemplify the style and concerns of the series. The programme starts with WHO broadcast reports of yellow fever, influenza and polio outbreaks from different parts of the world. Over the series title the commentary begins, clarifying the technocratic intention behind its name:

> The World is Ours, or so we like to think. And why not? Haven't we worked to make it ours? Haven't we multiplied and made fruitful the earth? Haven't we subdued its creatures from the largest to the smallest? Aren't we secure in our mastery? We like to think so, but this flask denies us.

The flask contains influenza virus samples from Paris, brought in by aircraft to be analysed at the World Influenza Centre, Mill Hill. The commentary asks whether we are winning or losing the battle against viruses. Smallpox, Polio, Foot and Mouth, Rabies and Tobacco Mosaic Virus are briefly covered. Then, in a major section, the virologist Christopher Andrewes explains the nature of viruses. White-coated and sitting informally on a table, he speaks in a semi-formal lecturing style, from the Mill Hill Centre. Receiving the flask, he begins with a brief outline history of virology and explains the size of viruses by reference to one of his own hairs. We are shown electron micrographs of influenza, smallpox and polio viruses. A section explains immunity. This introduces the next major part, on the 1952 Copenhagen polio epidemic of 2,300 cases. A Danish commentator explains the surprise that a city that prided itself on its cleanliness should have such a major epidemic. In a sequence with music, pins in a city map indicating polio cases are intercut with scenes of telephonists and ambulances bringing cases to hospital. Control measures and therapeutics are explained. The account concludes by explaining the continuing necessity of rehabilitation.

In the next section, Anthony Payne of the WHO explains why polio, from being a rare condition, had produced widespread epidemics over the previous fifty years. He speaks directly to camera, evidently reading a script. He expounds the theory that polio is a disease of cleanliness, based on the understanding that children exposed to the virus under the age of five tend to have a minor illness, with more severe effects the older the victim is. There is

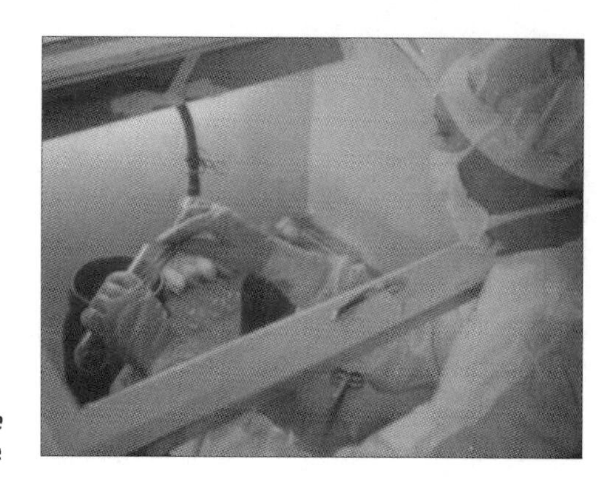

Fig 6.2 ***The World is Ours*: *The Invisible Enemy* – making polio vaccine**

a comparison with Britain a century before, using footage from John Monck's *Health for the Nation.* The engineering public health that washed-away the bacteria of cholera also removed the polio virus. So, there was less exposure, less infantile paralysis (the old name for the disease) and less immunity.

The commentator picks up a new topic: the 30-year search for a vaccine. We see the 1954 trial of the Salk polio vaccine. Payne reports it to be safe and reasonably effective, but with questions to be answered about the lengths of its effectiveness, and whether it would work in infants and against all strains. The commentator concludes: '…but the victims live on, reminding us that we must never rest in our fight against the viruses if the world is to be ours'. A short section gives the example of Yellow Fever. Once again, a historical and epidemiological account explains how the slave trade carried the disease to the Americas. We are told how hygiene measures controlled the mosquito vector, but that victory was short lived once it was discovered that South American monkeys were acting as a reservoir. A Yellow Fever vaccine had been developed, but the WHO was concentrating on strict quarantine measures to prevent the transfer of the disease to Asia.

A major section discusses influenza, with a historical prelude including the 1919 epidemic. This introduces Dr Alick Isaacs at Mill Hill, who explains how the many strains of the virus make it impossible to develop a single vaccine. Undaunted by the complexity of the science, he explains the typing of viruses by haemagglutination and the manufacture of vaccines. In the final major segment, the programme moves to India to discuss vaccination against smallpox. An Indian voice-over describes an outbreak. There is isolation of the affected child and an attempt at mass vaccination, first unsuccessful because of fear, then acceptance after many deaths. A final sequence takes the port of London (echoing the start of the programme at an airport) and the work of the Port Medical Officer, checking arriving ships. The final commentary, over swelling music, proclaims:

There are two battlefields: one is your body, the other is our world. And in this battle, the discipline is trust. Trust in the ship to signal for the medical officer, trust in the passengers and crew to take every precaution. Above all, trust in those who fight the invisible enemy every day in every corner of our world.

This echoes the scientific citizenship theme of the earlier documentaries, promoting acceptance of expertise, in this case in the shape of the WHO's rational programme and the work of biomedical scientists. In all respects this programme is a documentary film, not unlike a Shell Film Unit production, rather than an example of a specifically televisual genre. One clue to its membership of the established film mode is in the character of the speech; commentary is very important to how the information is conveyed. The concession to the television style is in the rather longer pieces to camera by the scientists but, with the exception of Andrewes' performance, even this speech is scripted, formal and fully grammatical, unlike the looser styles of speech in live television.

Despite the high hopes – on both sides – of Rotha as Head of the documentary department, it became clear within a few months that he was not going to produce the new genres that McGivern dreamt of. Elaine Bell suggests that he was averse to the culture of meetings and what he saw as opposition to the use of film (see 1986: 70). In December 1954, McGivern indicated a broader point about televisual genres when he commented on Rotha's draft of an article, *Television and the Future of Documentary* (published as Rotha 1955). 'I feel there is a little too much emphasis on film and ciné celluloid – and I would have been most interested in a section on the difficult subject of the use of "live" (studio) documentary as opposed to carefully prepared celluloid.'[70] Rotha's annual report for his first year outlines some of the issues:

Mr Rotha found it difficult to adapt himself to the BBC method as opposed to the method of the film industry. Broadly speaking, the film method is to set up a unit for each film (or film series) to be made. The BBC method is to provide permanent Servicing Departments to cater for all Output Departments ... These difficulties have led to a 'softening', not a strengthening, of documentary output which ... has fallen too low.[71]

The quantity of programmes produced was a continuing issue in this period, and there is more than a hint of criticism that perfectionism in filmmaking was preventing developments in live programming. Rotha was asked to leave the BBC when his second contracted year came to a close. The BBC took the opportunity to close the department down and to redeploy the staff to other areas, mainly in talks and drama. Some soon joined new commercial television companies (see Bell 1986: 80). But the decision not to renew Rotha's contract was at least as much about the failure to create a new form as it was about him as an individual. There was also a generational effect at work here. Rotha was ten years beyond his period of greatest success. His method was to make documentary films at the BBC, but Cecil McGivern and others were looking to define a new medium. Rotha's model of the documentary film was dead. It was succeeded in the television talks and OB departments by specifically televisual genres, some of which had been in development for several years already.

Conclusion

Scientific organisations had sought to control the BBC's coverage of science, but the mechanisms they proposed had not been accepted. The BBC's most senior staff had retained the arrangement under which contact with scientists that affected programme content was left as the responsibility of individual producers. As McGivern pressed for innovation in non-fiction modes, the Talks and Documentary departments responded in ways that built upon the fundamental assumptions of their genres. Talks either made technology the subject of light entertainment or presented the expert seeking to convey as much as possible of the content of the 'hard' sciences. These were the growing points for television science coverage. Documentary, reflecting the social concerns of the broad genre, represented doctors and scientists as the agents of enhanced social conditions, notably the prevention of disease. And yet, television executives remained unconvinced that an effective medium for television science had been found. George Barnes, Director of Television, wrote to McGivern in 1952, citing the success of *Special Enquiry*, which he said had 'a technique which is capable of almost infinite reproduction ... Have we got it in science? At any rate *the preoccupation of our scientific talks with one small aspect of science* does not lead me to think so' (emphasis added).[72] This awareness of the lack of a reliable format for science did not convert immediately into programme ideas, but we may see it as setting the tone for the next decade.

7 TELEVISION SCIENCE GENRES 1955–65

> 'Science is a natural for television … But what kind of science? How should it be presented on television? And at what level?' (McCloy 1963: 11)

In the nine years between Rotha's departure from the BBC and the launch of BBC2 with its distinctive science strand, *Horizon*, the landscape of science programme making at the BBC slowly altered. The institutional, technical, scientific and social environment changed, and programme makers responded with new kinds of science programmes, which built on the emerging Talks tradition that stressed basic science. Non-fiction television was important to the BBC; the Director General, Sir Ian Jacob, wrote in 1956 that

> The standing of the Television service in the country depends very largely on its treatment of current affairs, that is to say, on its handling of news, OBs and what we call talks, including documentaries. Even the mass audience are bound to be affected favourably if we make a really good job of this part of our output, but our national position depends a great deal upon our standing with that part of the nation which is responsible for and actively concerned with political, economic and scientific matters.[1]

It is significant that this most senior of the BBC's staff should pursue the public service obligation in this way, eliding the interests of the mass audience with its reputation with elites. At the same time, the BBC was once again the target of approaches from the scientific establishment, first in 1958 in the form of a delegation and then from 1960 in the context of the Pilkington Committee on the future of broadcasting.

Departments, genres and the BBC's representation of science, 1955–60

The plan for the production of non-fiction television at the BBC after 1955 was to divide the work of the Documentary Department between the Talks and Drama departments. This consolidated the division between documentary drama and all other genres, including magazine programmes, series and complete films, which came to be known as 'talks documentary'. But, whatever the internal conflicts and whatever the significance of technological and institutional factors had been in the first half of the decade, after 1955 the approach to science that increasingly came to be favoured by television stressed basic science and the authority of the scientist. The main difficulty under the new arrangements was in maintaining levels of output, especially in documentary-drama, and of new ideas on the 'talks' side.[2] Many fewer documentary-dramas were made as compared with the early 1950s, leaving medical, public health and other social themes to be represented in other genres.[3] Several factors may have been at play here. The Drama Department may have felt that the genre was being forced on them. At the same time, it has been argued that the BBC's purchase of Ealing Studios in 1949 eventually made films cheaper to produce, reducing the need for live performance (see Briggs 1979: 218–19; Scannell 1979: 102). Also, from soon after its introduction in 1958, videotape permitted high quality recordings to be made, which also reduced the requirement for dramas to be performed live for broadcast.

The Talks Department, given the role of becoming the BBC's main centre for non-fiction broadcasting after May 1955, produced a significant number of science programmes and series. However, from 1957 its producers found themselves in competition with a newly active Outside Broadcast department, where the producer Aubrey Singer took a strong interest in science programming. We may see here a parallel with the demarcation of appropriate genres between departments that had exercised the Documentary Department from the late 1940s. But, where those debates were essentially between live programmes and films, in the new competition both genres were televisual at their core, and the distinctions were between two kinds of live programming – studio-based and location-based.

The use of live television had produced a revolution in both the visual and the verbal presentation of science.[4] Especially for those in the talks 'lecture' style, television had the advantage over film that the simultaneous use of several cameras allowed the producer to cut in a continuous flow between showing the speaker and close-ups of experiments or demonstrations. In filmmaking, such an effect could only be achieved by shooting the different components individually and editing them together to create an illusion of continuous action (see McCloy 1963: 11). The verbal counterpart to this live vision-mixed aesthetic was a departure from the older model of scripted speech from all participants, including scientists, to unscripted – though doubtless rehearsed – and therefore looser, more informal, speech.[5]

Science programmes from the Talks Department from 1955

James McCloy, who transferred from schools broadcasts to Talks in 1955 (BBC staff lists), became its most prominent science producer.[6] His pro-

Fig 7.1 ***Frontiers of Science*****: one week after Sputnik, ex-Astronomer Royal Sir Harold Spencer Jones explains the trajectory of a satellite**

grammes had an emphasis on scientific discovery, technical innovation and technology's practical usefulness, in a mode that prominently featured scientists and technologists as television performers. Amongst the programmes he produced was *A Question of Science*, an early example of a format taken over from Independent Television (ITV). Broadcast in 1956 and 1957, this was a fortnightly 30-minute programme presented by Arthur Garratt. Following the tradition of *The Brains Trust*, viewers were invited to send in scientific questions, which were answered in the studio by a panel of scientists, including Peter Medawar and Joseph Rotblat (see Pound 1956). *Frontiers of Science*, broadcast irregularly from January 1956, promised to 'bring into the studio the men who are pushing back the frontiers of human knowledge, and will show the experiments they are doing in the course of their research' (McCloy 1956: 27).[7] A selection of programmes gives the flavour: the first, typical programme featured scientists from the Institute of Aviation Medicine at Farnborough. The edition screened on 18 June, to commemorate William Harvey's *De Motu Cordis*, combined history with new techniques in cardiac diagnosis and surgery. The programme broadcast in October 1957 was provoked by the launch of Sputnik just a week before and featured explanation from chosen scientists of orbiting satellites, the design of rocket engines, the physiological issues of life in space – including an experimental space suit – and Patrick Moore advocating manned space travel.

McCloy proposed a science magazine programme early in 1958, arguing that without such a format it was impossible to cover medium-sized stories, gain sufficient coverage across science, technology and medicine, build an audience for science or respond with sufficiently detailed scientific coverage to relevant news stories, such as the surprise (and subsequently disproved) achievement of nuclear fusion in the Zeta reactor at Harwell. Both he and his superior, Grace Wyndham Goldie, Assistant Head of the Department, were concerned with competition from Singer and OB; she looked to McCloy to rectify this: 'I think departmentally we are looking ill-informed in the scientific sphere and I should like to discuss with you how we can improve this situation.'[8] Goldie made the case for the series on the grounds that the total quantity of science on television was insufficient in relation to both its importance and its popularity.[9] At issue were the specifically *scientific* aspects of news stories; in existing programmes such as *News Conference*, *Panorama*, or *Tonight*, references to science were 'in terms of news interest: personality interest: or the interest of the social and political implications which are involved. There is no place in which the *purely scientific* aspects are represented or considered' (emphasis added).[10] We may note that this was very different from the older concentration on science as the means to solve human problems.

The outcome of these proposals was *Science is News*, broadcast from October that year as a fortnightly half-hour programme, listed with the rubric 'television reports on discoveries in Science, Medicine, and Industry which are changing our world' (Anon. 1958: 19). McCloy's justification was that television 'is uniquely qualified to report on science because the viewer can see for himself what all the words are about' (1958b: 9). Presented by Roy Bradford (and later David Attenborough), it typically featured three items per programme on science (for example, bacteria in space; the life cycle of the locust), medicine (transplantation of ova; research on the common cold), engineering (channel cable for electrical power; the use of diamonds in industry; satellites) and issues (the psychology, physiology and physics of noise).[11] Within a year, however, the strain of producing this programme was proving too great. Goldie reported to her superior Kenneth Adam, who had taken over from McGivern:[12] 'We have failed to solve the problem of adequate scientific staffing, and although *Science is News* was increasingly authoritative and successful it had to be rested because of the overwork of its small scientific staff.'[13] That production staff was comprised essentially of McCloy.

Other examples of Talks Department output included, from April 1957, *The Sky at Night*, produced for many years by Paul Johnstone. This programme can be seen as exemplary of two traditions; it brought together the Talks tradition of the expert speaker (Patrick Moore) with the culture of amateur science. Moore himself had no formal training but, self-taught, became a noted expert in astronomy (see Levy 1997). This programme, like *Secrets of Nature* a quarter of a century before, had a mode of address that acted to encourage viewers to do amateur science, rather than simply to learn about it. The *Radio Times* entry for the first programme printed a star map, with the comment 'anyone can find interest and excitement in the night sky, if he knows what to look for' (Anon. 1957: 4). Astronomy, like the natural historical pursuits discussed in chapter one, has had a significant amateur following from at least the nineteenth century. At the other end of the scale was *The Hurt Mind*, broadcast in early 1957, which was based on preliminary researches into public attitudes to mental illness by the BBC Audience Research Department. It was produced by Andrew Miller-Jones, and presented by the MP Christopher Mayhew. This series, with programmes on mental hospitals, the causes of mental illness, psychotherapy and physical treatments, concluded with a studio discussion. In spring 1958, McCloy also produced a six-part series on human evolution under the title *Five Hundred Million Years* (1958a: 3).

From 1955, the Talks Department had largely ceased to make documentary films on scientific or medical themes on the scale of Rotha and Swallow's *The World is Ours*. One exception was *On Call to a Nation* (broadcast on 22 October 1958), which fell into the category of '"Major" or "sledgehammer" 60 min. documentary programmes'.[14] One reason for the rarity of such programmes may have been the investment required; this film took six months to make (Cawston 1958: 9). Written and produced by Richard Cawston, it was a 75-minute examination of the state of the NHS ten years after its foundation. At the beginning, the viewer sees patients queuing to see a GP. This is used as a structural device; the medical requirement of each patient introduces a specific aspect of the service. On each, there is a sequence of exposition voiced by the narrator Colin Wills, followed by the separate opinions of a selection of doctors in different types of practice, differentiated by whether they are urban or rural, serve large communities or small, and operate in partnerships or as group practices. Patients' voices are heard, but the narrator and doctors predominate. Using this structure and much location filming, the viewer is shown: the 'capitation system' as a basis for funding general practitioner services; prescriptions and mechanisms to control the costs of drugs; dental, ophthalmic, public health, pathology and diagnostic services; hospital in- and out-patients, and organisation; the connection be-

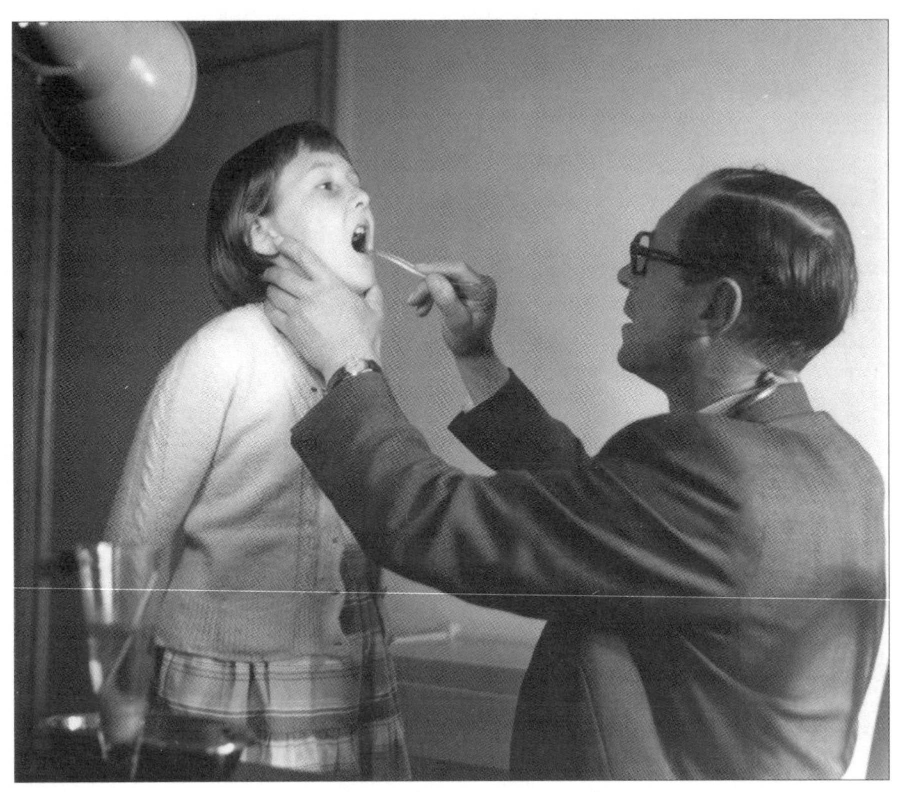

Fig 7.2 ***On Call to a Nation*****: the film introduced its different sections using the device of a sequence of patients visiting their G.P.**

tween scientific advance and the medical services; and the politics of costs. In formal terms it was very different, despite the fact that it was shot on film, from documentaries made for projection. Although closest in coverage to John Monck's *Health for the Nation*, this programme was explicitly neither a historical account nor a 'portrait' of the NHS; rather it was a current affairs report with the production values of a film. Unlike *The World is Ours*, this conveyed opinion rather than a documentary description. The difference is most evident in the style of speech; there is no reporter here to be the viewer's proxy, but the doctors speak unscripted and informal answers to what were quite clearly very specific questions as they provide the film's detailed analysis and critique of the state of the service.

Science programmes from the Outside Broadcast Department from 1955

In the second half of the 1950s, BBC managers encouraged producers to make new types of OB programmes, partially because the quantity of OB equipment in the regions had been increased to cover sports fixtures. As these mainly took place at weekends, equipment was underused on weekdays.[15] In 1954 for the first time, the department was listed as a production rather than a facilities department (BBC Staff Lists). It began to concentrate more on what participants described as 'built OB' programmes, that is those that did not merely transmit existing events into peoples' homes, but treated real venues, with all the authenticity given by location, as television studios, researching the scenarios that made such venues the natural sites for television. The producer Aubrey Singer, who had originally joined the OB department in 1949 but had taken positions in Scotland and the BBC's New York office before re-joining in 1953, became a crucial player in the development of science television within this department (see Nichols 1997: 63). Singer himself dated the OB's 'first incursion' into science to 1957 (1966: 3).

His first notable science venture was the high profile and ambitious programme, *The Restless Sphere: The Story of the International Geophysical Year (IGY)*, broadcast in June 1957. This was both big science and big television. The IGY involved international scientific collaboration over 18 months in standardised measurements of the characteristics of the interior of the earth, its surface and the atmosphere. With a commentary and links by Richard Dimbleby, the programme featured narration from the Duke of Edinburgh. It included, from around the world, both live OB sections and telecined film sequences. Singer described it as 'the greatest international coverage ever attempted by the BBC' and alluded to the year of meetings between the BBC and the Royal Society that had been necessary to create this joint project (1957a: 5).

But, amongst scientific and technical subjects, the breakthrough series produced by the OB Department was *Eye on Research* (1957–61), which took live outside broadcast cameras to various research establishments, and *Your Life in Their Hands* (1957–64), which went into hospitals (see Lawrence 1990 and Loughlin 2000). The approach of *Eye on Research* was journalistic; the billings in the *Radio Times* featured prominently the reporter – Robert Reid (from *Special Enquiry*) in the first series, then Raymond Baxter. Like *Special Enquiry*, this marked a departure from the dominant modes in documentary films. The films that Rotha had characterised as 'journalist' in the 1930s had not used a reporter to be the viewer's proxy in relation to their subject matter. *Housing Problems*, for example, framed the participants' testimony within 'voice of God' narration and expert opinion. In *Enough to Eat?*, Huxley was not the viewer's representative, but the public scientist, the voice of expertise (see Corner 1991b).

Singer proposed the production of a series about research in August 1957, shortly after completing *The Restless Sphere*. He stated that the pro-

Fig 7.3 *Eye on Research*: Raymond Baxter explains the world of the electron microscopist

gramme would be 'aimed at a "middle of the road" audience, showing what the backroom boys of science, industry and the arts, are doing in the way of research'.[16] At this proposal, the contested nature of science coverage at the BBC became clear. James McCloy, at that time producing *Frontiers of Science* for Talks, wrote to Singer about differentiating the new series from his own, proposing that it be limited to industrial research. The televisual genres typical of the two departments were crucial to the argument; McCloy said that 'since I cannot do satisfactory studio programmes on these industrial projects in the studio it seems a pity to waste OBs on pure science and medicine that really demand a studio treatment'.[17] Singer was unmoved by such approaches; when it was followed up by Goldie, McCloy's head of department, suggesting a conversation so as to avoid departmental clashes, he scribbled on the bottom of the memo, quoting Lewis Carroll, 'Beware this jub jub bird and shun this frumious bandersnatch.'[18] In deciding Talks Department programmes, competition with Singer was frequently mentioned, as for example when McCloy wanted to press ahead with a cancer story for *Frontiers of Science* ahead of an *Eye on Research* on a closely related topic.[19]

Each week *Eye on Research* reported on a particular field of science by interviewing scientists in their laboratories. The first, ten-part series included visits to the Royal Aeronautical Establishment for a programme on aerodynamics testing; to the Burden Neurological Institute in Bristol to speak to William Grey Walter on 'the mysteries of the human brain'; and to Manchester University to discuss computers. The last programme in the series combined feeds from several studios and film inserts for a programme about DNA: 'The story of how physicists, chemists and biologists have combined forces to produce the biggest biological breakthrough in the last fifty years.' The series was produced by a small core team of Singer as series producer; researcher and writer Gordon Rattray Taylor; and a roster of half a dozen producers and directors, including Bill Wright and Philip Daly, in addition to the reporter on-screen. The picture of science presented by Singer was upbeat and positivistic, linking novel technologies back to the laboratory:

> The age of science has been with us a long time ... The major scientific advances we read about, such as satellites, atomic energy and automation, are the culmination of many years of thought and experiment in laboratories all round the world. The long periods of painstaking exploration that lie between the bold announcements get little mention. (1957b: 7)

The programme emphasised this aspect of scientific endeavour, the science in progress, not quite Kuhnian 'normal science' (Kuhn 1962: *passim*), but perhaps its positivistic inversion, the science that would lead to the scientific advances we shall read about. The definition of scientific research was deliberately wide:

> It ranges from atomic energy to the preservation of fish; from the workings of the human brain to studies of the properties of cement. Research in this country alone embraces the whole universe, from the distant galaxies to the particles within the atom itself. There can be few things today that are not attracting the scientist's interest. (Singer 1957b: 7)

The tone was conventional modernistic techno-enthusiasm: 'As technological advance continues, its speed is likely to increase.' Unlike Rotha's view of science stressing those aspects that work for human welfare, this was a wide-eyed, excited, account of scientific research, in the mode of Oliphant's concentration on representing science 'more as natural philosophy, as a way of life and a culture in its own right'.

The selection of themes started with making, via conversations with scientists, a balanced list of subjects which 'must be reasonably popular with viewers, and at the same time be talking points in the world of science'.[20] This idealised list was narrowed down by the availability of OB units across the country; by the wish to balance university, government and private industrial laboratories and a concern not to cover a similar subject in consecutive programmes. The final balance across the series was towards the applied sciences; in more than forty programmes over seven series, *Eye on Research* contained rather few theoretical subjects. Gordon Rattray Taylor outlined the problem on a visit to the physiologist Lord Adrian:

> I explained the difficulty of achieving a balance in the series as a whole, due to (a) the more photogenic quality of applied science, which led to an over-emphasis on applied work; and (b) the difficulty of finding ways of presenting work in chemistry and the social sciences, resulting in an over-stressing of physics, particularly.[21]

The exceptions stand out: *The Particle Hunters* (24 February 1959), a visit to CERN; Dr Nicholas Kurti on *Absolute Zero* (17 May 1960) and the DNA programme. Singer's confidence and ambitions for the programme were clear from the way that, from the second series onwards, he dropped his voluntary exclusion of medical subjects, undertaken in the first series to avoid clashes with *Your Life in Their Hands.*[22]

Smaller Than Life: Story of the Virus, broadcast in September 1959 as the first programme in the second series, was concerned with research into the nature of viruses. Presented by Raymond Baxter from the Cavendish Laboratory, Cambridge, it relies on him both to introduce and conclude the programme, and to summarise the contributions of the several scientists who explain their work. Typical of the ambition of the programme it includes not just interviews from two laboratories at the Cavendish, but also a live OB from the International Exhibition in Brussels where Aaron Klug speaks of viral structures, including bacteriophages; and from the National Institute of Medical Research, Mill Hill, where Alick Isaacs explains about the new discovery of interferon. There is also a film sequence from Germany on tobacco mosaic virus. The programme uses several animated diagrams, and sets great store by the excitement of showing live images from light and electron microscopes. The impression given is of a report from the front.

Scientific participants responded enthusiastically to the series; for example, an official at the government's Department of Scientific and Industrial Research (DSIR) wrote to Singer about the National Physical Laboratory programme:

> It has aroused a good deal of interest in our research stations and has succeeded in 'selling' television as a means of putting across scientific information in a reasonably interesting way ... Your series has shown that much

> greater interest is aroused by visiting the scientist in his natural habitat rather than by bringing him, together with a few bits of his equipment, to the studio. You can count on enthusiastic co-operation from DSIR in any future series.[23]

The programme received good audience responses, with 'appreciation indices' around 71.[24]

Eye on Research shows how use of the Outside Broadcast mode affected how science was represented to the public. There was something of a fetishisation of the simultaneity of the medium, in which the fact that the broadcast was live from laboratories reinforced the sense of being up-to-the-minute in scientific terms. Producers were aware of this factor; as Norman Swallow commented of the sister programme, *Your Life in Their Hands*, 'to transmit such a sequence live is infinitely more effective than to pre-record it, for there is always an added sense of occasion in being present when something dramatic is actually happening. To be allowed to watch something which took place yesterday or last week is a poor substitute' (1966: 148). Yet there was a difference: in the medical series the viewer saw, for example, an operation on a real patient's liver; in the science series they were generally shown the working environment of scientists and technologists and equipment was demonstrated. Science was explained by articulate scientists, but the audience did not see science in action as they saw surgery in progress.

Issues of comprehensibility

One critical response to these programmes identified an important corollary of their chosen approach to science. The *Listener*'s reviewer Hilary Corke wrote:

> The scientifically-educated find it hard to realise the deficiencies of those who are not – they do not set their sights low enough: in this programme on radio stars, and the use of radio as opposed to optical telescopes, the relationship between light and radio waves was never stated, and the matter must have remained darkened for many. (1960b: 680)

This was the issue of comprehensibility that had also concerned the Anderson Committee in 1950. It is a safe generalisation to say that this had not troubled filmmakers in previous decades; it was generally considered to be the task of film technique to convey what was required. But, to a significant extent, this problem was created by the insistence on conveying the scope and content of modern science. Another reviewer complained that 'scientists naturally get carried away by their own work and want to explain it fully; simplification often gets lost by the way' (Gransden 1958: 704). The issue was also addressed in a *Listener* article at this time by Lawrence Bragg, then Director of the Davy-Faraday Laboratory at the Royal Institution (RI). By virtue of this position and also his involvement on the BBC's General Advisory Council a decade earlier, Bragg had visibly become a 'public scientist' in Frank Turner's terms. His article appeared just after his series of six lectures from the Royal Institution on *The Nature of Things* in November and December 1959.[25] This was a series of televised lectures, produced by Philip Daly for the OB department, which could be said to fit Robert Barr's definition, from 1951, of a 'talk' as 'expert ... information ... conveyed directly from the authority to the viewer' (see chapter six). Here Bragg, a senior scientist, explained principles of physics and chemistry in 15-minute programmes that featured lecture demonstrations conducted by Bill Coates of the Royal Institution. Bragg's *Listener* piece reflected on the different potential modes of science broadcast. Recalling the Anderson Committee-era research on comprehensibility he suggested three types of scientific broadcast, with differing levels of address and difficulty. The first, which he named 'The Achievements of Modern Science', referred to 'something man could not do before but can do now, owing to the advance of science, [such as] moon-rockets, nuclear power stations, a new wonder-drug'. Generally, he argued, it is impossible to explain the scientific principles involved in such things because they are too complicated: 'if the aim is to give the man in the street some idea of what science is, and to convey the thrill of understanding its explanations, such talks in general do not provide a good opportunity' (1960: 75). This might describe *Frontiers of Science*, whilst his second category, 'Recent Advances in Pure Science' could almost be the rubric of *Eye on Research*: 'News gets round that here and there, in some laboratory, a researcher is doing work of fundamental importance on the frontiers of science which may lead to great things.' He asks a question and supplies his answer:

> Can it be explained to 'the man in the street' what he is trying to do? Here the answer, it seems to me, is definitely 'No, it can't', except on the most broad and general lines [because] any real explanation involves using technical language ... Further, such pioneer investigations on the scientific frontiers are far removed from any every-day experience, and also far removed at this early stage from any practical application. (1960: 2)

It was his third category, 'The Science of Everyday Things' – which, unsurprisingly, described his series – that he felt 'offers the real opportunity to interest the listener in scientific explanation'. As he enthused, 'We all use many new materials, and many new mechanical or electrical devices, which have been made possible by the advance of scientific knowledge. Most people like to know how things work and get pleasure from a clear explanation.' Bragg emphasised the scientist's communication and largely ignored the power of the medium although, discussing his RI lectures, he conceded that 'the demonstrations ... gained greatly from the clever way in which the BBC experts took their shots' (ibid.). So here we have a significant difference of opinion about science television: on one side, *Eye on Research*, pursuing the scientist to the bench; on a second, McCloy presenting articulate scientists explaining rocket motors; and on a third, the grandiloquent science communicator conveying basic science via information on how things work. It is notable that, within his three categories – of scientific achievements, pure science and the 'how it works' of everyday things – there was nothing that looked at all like the science in the service of social welfare that we witnessed in earlier chapters, and which had come to be associated with the political left. There is a hint here of Cold War attitudes affecting what types of science were to be represented. Singer, McCloy and Bragg may have differed on what the precise focus of television science should be, but they all emphasised the transmission of the content of modern science, rather than its effects.

Reacting to scientific pressure, 1958–64

Despite all this activity in science broadcasting, the scientific organisations were not content, and in September 1958 the Royal Society and British Association once again approached the BBC. A meeting was held between the Director General Ian Jacob and a delegation comprised of the President of the Royal Society Cyril Hinshelwood, President of the BA Sir Alexander Fleck and, from the government's Chairman of the Advisory Council on Science Policy, Sir Alexander Todd.[26] This meeting was contemporary with the launch of the second series of *Eye on Research*, only 15 months after *The Restless Sphere* and a month before *Science is News*.[27] The approach was initially different from earlier examples in that the delegation pressed for the establishment of a specialist BBC science department, which they felt would place science on an equal footing with drama or religion.[28] They linked this not simply to a greater quantity of scientific broadcasts, but wanted 'an organised programme of popular science broadcasting ... so planned as to cover the whole field ... over a given period'. More ambitious than the Anderson Committee, they sought an audience not only of scientists or even those with some scientific knowledge, but the 'broad mass of the people so as to interest them in scientific development and to present science as vital to the well-being of our contemporary world'.[29] In short, they were looking to television to educate the general public systematically in science.

Jacob, consistent with his sensitivity to élite groups, took these suggestions seriously. He started a series of internal discussions with the producers of scientific programmes from Talks, OB, radio and the regions. Not dismissing the suggestion out of hand, he asked producers to consider whether a science department was practicable or desirable or whether a liaison officer might prove a good alternative.[30] Kenneth Adam primed Singer and McCloy to write papers describing how their programmes were planned, what outside sources they went to, and 'how you ensure not only that programmes are authoritative but that the subjects chosen are *what scientists themselves would regard as being both important and suitable*' (emphasis added).[31] The responses reveal both a fierce commitment to producers' independence and the mechanisms by which the representation of science came, in any case, to coincide with the views of élite scientists.

Both producers emphasised their access to senior scientists, McCloy stressing the advantage that being 'a trained scientist' gave him. Singer emphasised élite contacts: 'We generally start by visiting a group of scientists who are in touch with a wide field of research; people like Professor Thompson at Oxford, Sir Charles Harrington at the Medical Research Council and Dr David Martin at the Royal Society ... In these conversations we learn what is of topical interest to the scientist and we then make detailed investigations.'[32] McCloy expressed his respectfulness towards science directly, whilst asserting the importance of interesting the viewer; 'whatever the showmanship involved in presentation, the programme aims at being entirely responsible in its treatment of science. It must be responsible not only in question of fact but also in selection and emphasis, and *earn the good will of the scientific profession*' (emphasis added).[33] However, in the selection criteria for stories there was a significant difference between the two departments. Singer stressed that 'the determining factor in the selection of scientific subjects for television was their suitability for the medium. Some branches of science were not good television material.'[34] McCloy stressed the issue of comprehensibility: 'There were aspects of science which were incommunicable to the mass audience, however hard one might strive to present them in simple terms. Some branches of science defied simplification.'[35]

Neither supported Jacobs' compromise of appointing a coordinator; Singer could not see how the BBC would find a sufficiently senior scientist willing to do the liaison job for the money; McCloy argued more philosophically that 'science was not an entity and that coordination would therefore prove extremely difficult'.[36] On the issue of quantity, Kenneth Adam conceded that the television service was producing fewer science programmes than might seem desirable, but he argued that this was partially because of the shortage of suitable production staff. McCloy countered that, as science had become a vocation, it was difficult to persuade the newly graduated scientist to turn to television production.[37] At the subsequent Board of Management, the DG concluded that neither a specialist department nor an individual coordinator was appropriate; he proposed instead greater departmental liaison, a scientific approach to programmes and the recruitment of young scientists as trainee producers.[38]

By the time a sub-committee chaired by Jacob and comprised mainly of producers reached their conclusions the following May, a substantial paper had been prepared and discussed at the General Advisory Council at the request of one of its members, Sir Hugh Linstead.[39] Here pressure was again exerted to put a senior individual in charge of all science broadcasting. The DG reported that senior figures in science, including Hinshelwood, wanted more for scientists than day-to-day collaboration with science producers and that they could not see why there was no BBC head of science when there were senior figures in charge of areas including education and religion. He suggested freeing the senior producers in radio and television science from programme making so that they could devote themselves more to liaison and suggested that the different science producers from television and radio should meet occasionally for the sake of coordination.[40] A review of the situation in autumn 1960 revealed that no such meetings had taken place.[41]

Despite the intensity of the discussion and the BBC's successful defence of its producers' autonomy from scientific control, the actual practice of television led to programmes that incorporated the view of science promoted by its élites even whilst it repulsed their control at a higher level. In 1960 both the Talks and the OB departments produced series to mark the tercentenary of the Royal Society. McCloy's *Life Before Birth* was a five-part series on human reproduction (see Briggs 1995: 466), whilst Aubrey Singer devoted the fifth series of eight programmes of *Eye on Research* to the work of Fellows of the Royal Society.[42] In July, the BBC also broadcast the Queen's formal opening of the celebrations from the Royal Albert Hall.[43] Although the aims of the delegation were not fulfilled, the Royal Society, and science and technology more generally, were doing well for television coverage. Furthermore, whatever the insistence of producers that it was their prerogative to select scientific subjects that worked in televisual terms, the picture of science represented veered strongly to the Oliphant view of the subject as 'a way of life and a culture in its own right' that was entirely congenial to scientific élites. For example, David Martin from the Royal Society had written to Philip Day on the subject of *Eye on Research*:

I would like to put on record how very valuable I have found [these programmes] to be, not only from the point of view of the success of our Tercentenary Celebrations, but also in the whole development of the presentation of science on television. I have heard many golden opinions about the series from Fellows.[44]

This tendency continued in the following decade, even to the extent of broadcasting structured courses of lectures, including a repeat of the Bragg series and a sequence of ten by Herman Bondi on relativity, as part of an experiment in adult education, broadcast on Saturday mornings (see McCloy 1963: 11).

Changing representational conventions

Science International, produced by Singer and broadcast in December 1959, had extended the approach of *Eye on Research* to the global scale, with two programmes: *What is Life?* and *The Last Scourge* (on cancer). A review of these programmes by the documentarist Michael Clarke (who had started with Rotha) reveals how filmmakers and television producers were beginning to have rather different assumptions about the visual representation of science. Describing them as 'probably the longest, most concentrated and most complex television programmes on science so far devised, in Britain or elsewhere' (Clarke 1959: 303), he proceeded to consider communication versus spectacle, the effectiveness of presenters, the fetish of topicality, and comprehensibility. He drew a distinction between a programme that sought 'to help people understand some aspect of science and those that set out to merely impress them with the scientist's own erudition' (ibid.), which he defined as 'spectacle', also present in the shape of a giant model cell on set. He believed these programmes to have veered from explanation to 'spectacle'. He argued that the use of presenters – in this case Raymond Baxter – prevented, rather than enabled, effective communication by the featured scientists, especially the molecular biologist Professor Michael Swann. He felt this betrayed an attitude that scientists constituted a race apart: 'We have to deduce that the BBC used Baxter as an interpreter … because it wanted to preserve the idea of science as a closed circle, a secular Church that only the few can enter' (1959: 304). Most of the review was concerned with issues of comprehensibility. Clarke started positively: 'For once, the directors did not quail at the difficulty of modern scientific concepts, or suppress all that could not be pre-digested into meaningless pap' (1959: 303). However, he felt that, for viewers unfamiliar with the subject, 'the programmes were too difficult; and a whole hour proved too long. One felt that the producers had not had time to put themselves in the audience's position' (1959: 304). His view was also that the language, which was 'too technical' (ibid.), and the visual techniques – including the giant cell model, live electron micrographs and some of the film inserts – failed to explain the programmes' content. But, he argued that balancing Bragg's *Nature of Things* with these programmes on cell biology – 'about the most difficult subject they could have chosen' (1959: 305) – prevented the BBC from falling prey to the tendency of television science programmes' 'apparent obsession with the topical and the new' (ibid.). The filmmaker Clarke's view agreed in part with that of Bragg the scientist (whose *Listener* piece was published the following year). Both rejected the concentration on novelty (as exemplified by the output of both Talks and Outside Broadcast), but where the latter wished to seduce the ordinary viewer with the science behind everyday things, Clarke's preference for the medium was that it should render difficult modern science comprehensible. These were the views of a filmmaker with significant experience of scientific subjects, but not of the pressures of live television.[45] In sum, the core of his concern was that particular televisual techniques, including the use of an 'anchor man', models and electron microscopes, to his taste mystified science when his belief was that effective visual technique should, by contrast, explain. He believed in a cinematic clarity of explanation; elsewhere he asserted that 'clarity has its own high aesthetic value' in scientific filmmaking (Anstey *et al.* 1963: 8). This pointed to an ideal of 'demonstration', 'instructional' or 'factual' scientific film (in the old Scientific Film Association categories) that continued to be influential for classroom use (Alex Strasser's *Mirror in the Sky* from 1956 is one example), but television producers had not pursued this more academic style.

The Pilkington Committee and the regularisation of television science and advice

The Pilkington Committee on the future of broadcasting, established in 1960, provided another opportunity for scientific organisations to seek influence over broadcasters. The BBC was obliged to respond to the organisations quoted in the committee's report of June 1962, the Royal Society, the BA and the DSIR.[46] The Royal Society and BA had stressed 'the compelling need to present to the public the significance of science' and their wish that there should be a greater quantity of science and technology programmes. Once again they proposed that a scientist be appointed to the senior programme production staff of the BBC, 'to take charge of science programmes' and the establishment of a scientific programmes advisory committee.[47] Both sides acknowledged that the Royal Society, with Fleck now its Treasurer, was pursuing exactly the same criticisms as they had voiced in 1958.[48] The

DSIR's concerns had centred on a perceived under-representation of technology compared to science. R. D'A. Marriott, the Assistant Director of Sound Broadcasting, and Stuart Hood, Controller of Programmes, Television, were deputed to liaise with the scientific organisations and to 'clear up this recurrent problem of satisfying the scientific world about what is loosely called the coordination of science broadcasts'.[49] The tone of the internal debates and response was very much of tactical play to resist pressure, firstly by having discussions with the scientific organisations conducted by working producers, who could state that they were unable to make policy, rather than by the DG's own staff. The second element was to make concessions over issues that mattered less to the BBC's independence and so retain control where they wanted it most, namely in not being obliged to have a senior scientist imposed high within the BBC's structure. Accordingly, they deflected questions about the quantity of television programmes to the period after the establishment of BBC2 and conceded the formation of a Science Consultative Group. A brief quotation from the BBC's account of the key meeting at the Royal Society conveys the tone:

> We had no objection in principle to the formation of a Committee; our only concern was that if it were to exist it should be useful and help in the making of good science programmes, and we thought that the BBC would prefer to avoid creating another formally constituted advisory committee.[50]

The Committee, which first met in May 1964 under the chairmanship of Professor Alexander Haddow, with nominees from the Royal Society, BA and DSIR, became a long-term fixture of the BBC, meeting twice per year over several decades.

The more significant impact of the Pilkington Report for science television was its recommendation that the BBC launch a second television channel. It was the establishment of BBC2 that permitted the development of its coverage of science on television, limited as it had been up to this stage by the requirement that the single channel serve all interests. This major change entailed a substantial reorganisation of production departments that went live in February 1963. Under this, Aubrey Singer became Head of a new Features and Science Programmes department within the Outside Broadcasts section. From the evidence of *The Restless Sphere*, and even more *Science International*, television science broadcasters had converged on a style. The difference between a studio-based Talks programme with some film and OB inserts and an OB programme from a venue converted into a studio using similar inserts is unlikely to have been very obvious to viewers.

By 1965 Singer's department was, as he noted, '"the prime contractor" for the production of science programmes in BBC Television [and] the main focus of contact for any enquiries or suggestions about science on television' (1965: 25). Other producers and writers working there included Philip Daly and Gordon Rattray Taylor, who had made *Eye on Research* with him, and James McCloy. After BBC2 started to broadcast in 1964, their scope expanded considerably so that Singer could boast an output of science programmes exceeding one hundred on both channels in its first year (see McCloy 1963; Singer 1965; Briggs 1995: 467).

Horizon

The programme that was emblematic of the BBC's new commitment to science was *Horizon*, originally a monthly show, first broadcast in May 1964 as one of the original roster of BBC2 programmes. In the first discussions about this series its proponents wanted to make it a scientific analogue to the arts magazine show *Monitor*, which had been running since 1958:

> *Horizon* will attempt to present science as a culture, a field of human achievement and endeavour as lively, varied and as rewarding as any other. *Horizon* will recognise that any scientific discovery or idea is a personal creation stamped by the character of the scientist and his age ... *Horizon* will attempt to place new developments in science in their total context: personal, social, historical, political.[51]

Philip Daly stressed ideas: 'The intent always must be ideas and the problems associated with those ideas: never the straightforward teaching and demonstrating approach as an end in itself.'[52] After an unsuccessful pilot the producers were criticised by the BBC Chief of Programmes for making the show too derivative. This started the process which, from the beginning of actual broadcasts, led *Horizon* only rarely to cover more than one story per episode, and seeking to establish its identity not by having an anchor man, but by distinctive opening titles and music.[53] The production team, which included Daly and Ramsay Short (who had previously worked at the Shell Film Unit and had directed their Royal Society tercentenary film), began work on a final definition for the first series in January 1964. This included a definitive move towards single-theme programmes, even where several items were unified under one heading.[54] The philosophy was accepted in internal discussions and became the basis of the *Radio Times* introductory article, where Daly explained:

> The aim of *Horizon* is to provide a platform from which some of the

world's greatest scientists and philosophers can communicate their curiosity, observations, and reflections, and infuse into our common knowledge their changing views on the universe ... We shall do this by presenting science not as a series of isolated discoveries but as a continuing growth of thought, a philosophy which is an essential part of our twentieth century culture. (1964: 8)

Unwittingly, perhaps, the BBC had ended up with a flagship programme that would have made perfect sense to Oliphant, with his promotion of scientific broadcasts 'more as natural philosophy, as a way of life and a culture in its own right'. However, the previous year's ambition to show its 'total context: personal, social, historical, political' did survive. Its first two series show that, as it established itself and soon doubled its frequency,[55] it retained a significant breadth. The programmes from 1964 give an idea of the range and how they realised the ambitions for this show on the culture of science. The first episode (2 May 1964) was a profile of Buckminster Fuller, inventor of the geodesic dome. This clearly shows how the ambition to transfer the arts programme format to the scientific sphere worked. In this portrait, Fuller is treated as a creative figure worthy of respect, with juvenile sketches accorded the respect of a latter-day Leonardo. Much of the programme is taken up with his lecture to students at the Architectural Association. For long periods the film camera stays immobile, recording his explanation of the strength of different structural components and speaking about his philosophy. In later sections he is seen discussing the structure of viruses with Aaron Klug at the Laboratory of Molecular Biology in Cambridge and the electron microscopist Robert Hall.

The second *Horizon* (30 May 1964) picked up the debate on chemicals in the environment; the third, *A Candle to Nature* (27 June 1964), visited the chemistry of candles in a comparison between a lecture by Michael Faraday from 1860 and a modern one by the chemist George Porter. The next, *Strangeness Minus Three* (25 July 1964) was on particle physics, featuring the physicist Richard Feynman. This was followed by a programme, *The Air of Silence* (22 August 1964), summarising medical research at the National Institute of Medical Research, featuring many interviews with scientists, including Peter Medawar. *The Knowledge Explosion* (21 September 1964) was a discussion between Arthur C. Clarke, Derek de Solla Price and Nigel Balchin about the growth of science in the twentieth century. It featured Albert Einstein, atom bombs, nuclear reactors and particle accelerators, atomic clocks, electron microscopes, DNA, plastic and nylon manufacture and modes of transport including flight. *Tots and Quots and Woodgerie* (16 November 1964) re-staged a meeting of Zuckerman's dining club and considered a similar American

Fig 7.4 *Horizon* reproduces an older mode of science television: the chemist George Porter lectures on the science of candles

group. In a special edition they broadcast the pre-recorded self-obituary of J. B. S. Haldane. A Christmas show (14 December 1964) discussed scientific toys and entertainments.[56] From the beginning some episodes of *Horizon*, like the preceding efforts of the Talks Department, brought together substantial film sections with some live studio; in 1966 it settled down to being a series of filmed documentaries (see Silverstone 1985: 3). Programmes continued to be shot and edited on film into the 1990s.[57] The early broadcasts of *Horizon* supplied an element that had been rare in television science since Rotha's *The World is Ours*, that is, extended, crafted descriptive programmes about science, made ahead of broadcast. They were also mainly without the drama of live broadcast that had become such a fetish of those, such as McGivern, responsible for the creation of televisual genres.

Tomorrow's World

As Singer became established in his new role, he revisited some of the territory not just of his old department, but also of Talks. For example, in his first arguments about the need for the science magazine programme for BBC1, initially under the title *Modern Age*, launched in 1965 as *Tomorrow's*

World, he effectively repeated many of the arguments that James McCloy had made less than a decade before for *Science is News*. He cited governmental interest, public curiosity about expenditure on science and technology and the increase in popular scientific journalism, with the appearance of journals such as *New Scientist*. Writing only a month after Harold Wilson's 'white heat' speech to the Labour Party conference, he had concluded that 'these trends tend to confirm this important shift in current affairs. People are coming to believe, right or wrong, that in science and technology lies the way to the new frontier.'[58] Glyn Jones, its producer, picked up the argument:

> Our brief is to show new developments, to relate them to current affairs and explain them to the people. This is a plain man's magazine. Subjects, however difficult, must be crisply and simply explained ... Relevance is essential. We know viewers can be excited by spectacular developments in science and engineering. We want to capitalise not only on excitement but on the relevance our subjects have to the ways politicians act and to the way men and women live – and earn their livings. We are concerned with the source material for nearly all political, social and industrial decisions.[59]

This programme, a live broadcast with OB and film inserts like its predecessors *Frontiers of Science* and *Science is News*, was immediately popular with audiences. Its first programme, presented by Raymond Baxter and featuring two extended items on new medical technologies and on the potential applications of computing to business, received an appreciation index of eighty. One of the sample group enthused: 'I am eagerly looking forward to seeing more wonders of science presented in such a fascinating way.'[60]

Film versus television

Film and television producers debated the televisual representation of science in this period. At the 1963 meeting discussed in chapter five, at which Geoffrey Bell devoted his talk to 'factual cinema', James McCloy made a contribution specifically about his experience of television. He described a spectrum from televised science lectures in what, for the viewer 'is essentially an eavesdropping situation' (which might describe the Bragg lectures) to 'the opposite extreme [where] television can eliminate all personality and borrowing the traditional scientific film technique, present its subject in pictures, diagrams and demonstrations ... This is possible because the bulk of science is concerned with physical objects and their behaviour and not at all with people.' Both these extremes, we may note, could be captured under the SFA categories as 'instructional', 'demonstration' and 'descriptive' programmes.

Between these two didactic extremes for him fell 'all the television techniques evolved by television itself in documentary and feature programmes, techniques which can enrich a programme with all the intimacy so peculiar to this medium'. These included the immediacy and risk of live broadcasts and, especially, the personality factor, either of presenter or carefully selected articulate scientist. With this came more informal modes of speech than were found in the structured commentaries of older documentaries, although it must be conceded that this informality was often still as formal as a public lecture. McCloy once again addressed the issue of comprehensibility. He quoted an argument Hinshelwood had made about the supposed inherent difficulty of science:

> It is very difficult for any but experts to understand what is going on, not as so many non-scientific people are fond of asserting, because men of science are incapable of expressing themselves clearly, or unwilling to try, but for the simple reason that many aspects of the subject are of very great inherent difficulty. They are based on unfamiliar conceptions, often developed by advanced mathematical reasoning and sometimes expressed in a complex and abstract symbolism. (Quoted in McCloy 1963: 11)

In this sense McCloy and those who shared his views had produced something of a paradox; a circular, self-denying ordinance: to stress a supposedly true picture of fundamental scientific research which was considered incapable of communication to a mass public. Here, the argument of the documentarist Michael Clarke, as we have seen, was significantly different; that abstruse subjects could be made comprehensible by film (or television) technique.[61] McCloy spoke of two audiences: one 'with some background in science who are prepared to make a sustained effort', giving again the example of Hermann Bondi's ten-part series on relativity. By contrast, to reach the mass audience was simply a matter of showing 'the kind of thing the scientists are doing without attempting to go very deeply into the scientific principles involved', which was linked, for him, to showing the audience 'the scientist as a person' (ibid.). For all its paradoxical nature, McCloy's position had the interesting side effect of sustaining an élite place for scientific expertise above and beyond the public.

Aubrey Singer's perspective was different. In a public lecture in February 1966 he articulated his, and the BBC's, view of science broadcasting. His argument proceeded from the climate of public opinion on science to the types of science television the BBC should produce. Citing the dropping of the atomic bomb as the signal event that revealed the power and significance of science to the public, he listed three other scientific revolutions: in computing, with its adjunct information theory; in space exploration, launched by Sputnik; and molecular biology.

> Those of us engaged in broadcasting science to a general audience are forced to frame our attitudes in the light of this world we see around us. In doing this we have to take a stand somewhere … We place ourselves so that the inbuilt vested interests can be viewed as objectively as our own unconscious leanings of background and upbringing will allow. To this end, as a foundation to our policy, we have firmly decided *that the broadcasting of science shall be in the hands of broadcasters.* (1966: 8; emphasis in original)

Amongst the five principles of science television elucidated in this lecture, the most quoted was that 'the televising of science is a *process of television*, subject to the principles of programme structure, and the demands of dramatic form. Therefore, in taking programme decisions, priority must be given to the medium rather than scientific pedantry' (1966: 13; emphasis in original).[62] This comes as no surprise given all the interactions between scientists and the BBC in the previous two decades. However, it was a particularly confident BBC that could assert that 'the aim of scientific programming … is not necessarily the propagation of science, rather its aim is common with all broadcasting, an enrichment of the audience experience' (1966: 9). Difficulty of the subject matter was no particular barrier in Singer's view, on the assumption that 'the level of communication is between equals in intelligence: to an audience that is well disposed toward, but with no special knowledge of, the subject matter' (1966: 13).

This confidence extended to a statement about organised relationships with science; he publicly rejected the frequently requested mechanisms of central coordination and a central advisory committee, whilst warmly accepting the principle of the Science Consultative Group, as 'its suggestions, criticisms, and observations have proved extremely valuable'. But in case it was at all unclear, he stated 'in practice, however, most ideas come from producers, who, because they are working continually in the field, because they are creative and conscientious journalists, anticipate and fairly reflect what is of sufficient importance to make good television' (1966: 12–13).

Whether consciously or not, this emphasis on the role of the producer and on the primacy of science programming as a genre of television has continued to dominate the small-screen representation of science.

CODA: THE FATE OF GENRES IN TELEVISION SCIENCE SINCE 1965

In 1965, a decade after Rotha's departure from the BBC, television had left his model of scientific filmmaking far behind. It is beyond the scope of this book to address in detail the very wide range of television science programmes made beyond this date as television has produced so much science material that this account could not hope to be comprehensive; in 2007 the BBC programme catalogue listed 1,035 issues of *Horizon* and 1,375 editions of *Tomorrow's World* alone.[1] Even if this is only half the science television produced in Britain over four decades, it deserves its own focused studies. Accordingly, this generalising and speculative conclusion concentrates on reviewing the major genres whose gestation has been described.

Natural history

Natural history has long been a staple of television science. Alongside the early science programming described in the last two chapters were several examples of programmes representing the natural world. These continued some of the characteristics of genres from the beginnings of scientific cinema. Chapter one described the first popular natural history films, presented as a spectacle for the entertainment of the public. This genre developed at British Instructional and Gaumont-British Instructional. Julian Huxley's *Private Life of the Gannets* was a film made by a professional scientist but it stole many of the clothes of the amateur science tradition. After the war, films made for projection continued to treat natural historical subjects. From 1952 to 1961, Countryman Films produced several dozen films in the series *The World of Life*, a title later expanded to include the phrase *A Journal of the Outdoors*, covering a variety of animal and zoo-based stories. Another significant presence in this field has been Oxford Scientific Films, established in 1968 by Gerald Thompson, who began as an entomologist using cinematography within his research (see Anon. 2002). Many of Thompson's films were shown on the BBC series *Look*, which began in 1955 and ran for over a decade. Initially it specialised in presenting wildlife footage, often featuring the responsible filmmaker in discussion with the presenter, the ornithologist Peter Scott. Latterly it became more thematic, treating subjects such as extinction, and included live animals in the studio (see Gransden 1958). Scott expressed the programme's objective in terms of the importance of the conservation of animals, bringing to mind the language of Francis Martin Duncan – of providing 'the chance of enjoying animals, of delighting in them and wondering at their strange appearance or their strange behaviour and learning how to enjoy them more; because animals are like great music or poetry or fine pictures – the more you know them, the more enjoyable they become' (1958: 7).

Between 1955 and 1961 *Zoo Quest* presented stories about collecting animals for London Zoo, with the presenter David Attenborough on location in the habitats of the target species. With the foundation in 1957 of the BBC's Natural History Unit at Bristol, formalising an existing specialism there,[2] the popularity of films on plants and animals was institutionally recognised. As Gail Davies (2000b) argues, the Unit at Bristol not only represented but, in its programme making, also undertook work in natural history. Nature television, like the earlier films of Francis Martin Duncan and Frank Percy Smith, subsists on the technological revelation of living nature. This is as true of Attenborough's *The Private Life of Plants*, broadcast in 1995, as it is with *Battle of the Plants* (1926) from the *Secrets of Nature* series. In a sense, even the claim to the Bristol unit's filmmaking *being* as well as *representing* science is of a piece with how Percy Smith and his collaborators represented their films, in the books *Secrets of Nature* (1934) and *Cine-Biology* (1941).

Natural history television is a genre with a long history. This book's introduction suggests that genres may persist not because of conscious emulation but because, although the products of particular circumstances, they continue to have value to filmmakers. In this case, there is also some evidence of historical awareness from the example of *The Start of it All*, an issue of *Look* in December 1958 on the early wildlife cinematographer Cherry Kearton.

Films promoting scientific and technological modernity

The second chapter reviewed how non-fiction cinema responded to technological modernity. 'Interest' films had already discovered and were representing the transformative power of applied science and technology to the public well before John Grierson joined the Empire Marketing Board (EMB). This cinematically unsophisticated tradition has continued, a reasonable option for corporate film and video making. But chapter two also argued that the documentarists' self-conscious application of Modernist film technique to the representation of science, technology and medicine produced something qualitatively new; a Modernist cinematic form that knowingly celebrated technological modernity. Chapter five showed, with the example of *The Peaceful Revolution*, how the emblematic themes of the 1930s, shown in *The Face of Britain* or *We Live in Two Worlds*, were still doing good service in the 1960s. The technological transformation of societies and economies had renewed currency in the international world of the Cold War era. In 1961, as much as 1935, hydro-electric power was seen as a core technology of transformation but, modified by changing circumstances, in *The Peaceful Revolution* hydro-electricity was linked to a Third World development ideology.

Scientific and technological enthusiasm have been an inescapable component of both film and television representations. But uncomplicatedly 'gee-whiz' hymns to science and technology have been joined by accounts that are more critical. Notable amongst these have been moral and environmental critiques following the Aldermaston marches against nuclear weapons in 1958 and Rachel Carson's *Silent Spring* on the environmental impact of the insecticide DDT in 1961. Where editorial balance has been deemed important, it has come to seem obvious that celebration should be tempered with environmentalist sobriety, although for many years it was unusual to have scientists with opposing views debating in a single section of a programme (see Gardner & Young 1981: 177–8). *Horizon*'s second ever show, *Pesticides and Posterity* (30 May 1964), presented the contrasting views of the ecologist Dr Frank Fraser Darling and the industrial chemist Dr Eric Edson before concluding with a studio discussion. In *Man Made Lakes of Africa* (20 March 1972), the strand showed the environmental impact of the Volta and Aswan dams. Once again, the specific historical circumstances altered how science, technology and medicine were represented to the public. Especially after the 1973 oil crisis, television, particularly in current affairs mode, featured the environmental critique strongly.

The scientist's film

Chapter three showed how Julian Huxley developed a mode of furiously diverse public science in the 1930s that featured a significant commitment to the use of films in promoting the potential of science. *Enough to Eat?*, its emblematic film, made Huxley the spokesman for science's capability to diagnose and cure social problems.

Scientists have often considered public communication activities to diminish the seriousness with which popularisers' scientific work is seen (see Gregory & Miller 1998: 82–3). This opinion pulls in the opposite direction from the attempts of scientific organisations to participate in, and control, public communication. In Huxley's case, his time out of laboratory science led to him being passed over for the scientific adviser role that Dale fulfilled for the BBC in 1950. However, that should not be seen to diminish the impact that Huxley, presented as a scientist, had on the public face of science, even if it was as much via *The Brains Trust* as through more explicitly scientific vehicles. The scientist's film, in which the scientist appears as prominently as he did in the nutrition film, is, however, a comparatively rare beast. It is unusual to fuse the two professional roles – as 'anchor man' and simultaneously as voice of science. A partial comparison might be Robert Winston in those series remote from his primary professional expertise in reproductive biology, for example, 'How to Sleep Better' (2 February 2005). More usually, as television producers of all kinds have asserted across the period covered in chapters six and seven, they have valued scientists as the voice of science. This was explicitly present in Singer's 1966 lecture where he noted that 'wherever possible we use the scientist to tell his own story' (1966: 9). However, their contributions have most often been framed audibly by the persistent use of commentary (see Gardner & Young 1981: 178); visibly by the new professional role of television presenter, exemplified by Raymond Baxter; and more invisibly by the producers who structure the material.

But *Enough to Eat?* also set a precedent in its subject. Nutrition science, as the films discussed in chapters four and five demonstrate, has been an enduring fixture in the canon of films promoting science's capacity to diagnose social problems, and its asserted potential to deliver freedom from want. In formal terms, the most significant aspect of *Enough to Eat?* may have been that it linked science to documentary reportage. Although Rotha's films with John Orr chose instead the impressionistic and hortative modes, in television there has been an enduring association between reportage and science and this alternative impressionistic mode has only rarely resurfaced.

Scientific modernity, politics and documentaries

In Rotha's 1940s documentaries, especially the large-scale *World* films and *Land of Promise*, there was a contingent fusion of hortative cinematic technique with the social concern of particular planners and scientists. This was most visible with John Orr, but behind the scenes also many of the apostles of the social relations of science, including Julian Huxley, Peter Ritchie Calder, Edward Carter and Solly Zuckerman. Outwardly these films have much in common with the picture of science promoted by the Marxist J. D. Bernal in his *Social Function of Science* (1939). Ultimately however, they do not exemplify doctrinally pure Marxism, but are hymns to the power of scientific rationality. Also, like the Cripps model of planning, they insist on that rationality being willed by a (documentary-)informed electorate. If these films espouse revolution, the spirit is that of Orr, on behalf of the revolution that he believed science and rational planning could bring about in tackling deep-seated social problems. They were also predicated both on internationalist notions of world government that the United Nations agencies very soon showed themselves unable to deliver, and on a centralised national state at home that would fund the documentarists' ideology of persuasive information for citizenship. As chapter five described, Rotha's funding to make these passionate films had evaporated by the time he went to the BBC in 1953. Their extraordinary power and vehemence may, perhaps, have become an embarrassment for a socialist administration seeking to distance itself from colourful rhetoric and to prove its restraint as a responsible party of government (see Francis 1999: 152–63). Paradoxically for some of the best-made films in the canon, they are most time-locked in their relevance: they speak to the moment of their intended release. *Food Or Famine?* was in the same broad genre, but it was a quieter, more lyrical, less shocking film than *The World is Rich*, placing its emphasis on what science, quietly, could achieve.

The making and unmaking of television science genres

Aubrey Singer identified the atomic bomb as opening up public awareness of the size, cost and power of science. The petitioning by scientific organisations of the BBC in the quarter century from 1941 and especially from 1949 was second only in persistence to their approaches to government. This shows how organised science increasingly placed a high value on creating a positive public opinion of science. As the Cold War period developed, this concern with public representation hardened and reached a head at the time of the Pilkington Committee. In many ways, as this account has shown, the scientific élite need not have been concerned about how broadcasts showed science because, far from representing it incompetently or in a negative fashion, television producers were keen to convey a view of science that scientists approved of. This has often tended to what Oliphant had described as 'the great revolution wrought by the introduction of the experimental method, of the intellectual satisfaction and fun of science and of the scope and content of modern science' (see chapter six, note 11); broadcasters were keen that scientists should speak and be seen themselves as fulfilling this role.

Many of the genres of programme created in the first decade and a half of postwar television have had their successors in the decades since. The quiz-panel amateur inventiveness of *Inventor's Club* finds its echo in BBC1's short-lived *Best Inventions* (2002), just as the topical reporting of *Science is News* and *Frontiers of Science* relate to *Tomorrow's World* and *QED* (1982–99). The Science and Features Department gave science broadcasting more than

Fig 8.1 *Reality on the Rocks*: Ken Campbell's irreverent exploration of quantum physics

programmes; they also provided the apprenticeships that trained a diaspora of science filmmakers. Producers from here have included Patrick Uden, whose *Equinox* (from 1986) strand for Channel 4, whilst recognisably belonging to the *Horizon* family, focused especially on engineering. Uden Associates also produced the history of technology series, *White Heat* (1994). Windfall Films, another example, has made many episodes of *Horizon*, the quantum physics series *Reality on the Rocks* (1994) and *The Day the World Took Off* (2000), a long history of industrialisation. These series have escaped the conventions of *Horizon* and, even as they have sought ways to engage audiences with the content of science and technology, have introduced types of playfulness and irreverence that would have surprised their predecessors in the 1950s.

This broadening of genres was a distant effect of significant changes in society that affected science and representation alike. The culture that was concerned in the 1950s and 1960s with atomic weapons, nuclear power and DDT, with environmental causes of many kinds and protest against Vietnam, also produced new, more highly theorised, concerns about science, technology and society. Within the academy, one product of this analytical and critical spirit was the new discipline of science studies (see Porter 1990; Mayer 2003). One of its luminaries, Robert M. Young, with Carl Gardner, undertook a critical study of televisual representations of technical expertise (Gardner & Young 1981). Young's subsequent role as Chief Consultant and Series Editor of Channel 4's short-lived radical series, *Crucible: Science in Society* (1982–83) resulted from this study (see Young 1986: 4). With its pedigree in science studies, it aimed to introduce a particular critical account of the topic into television science, 'to contribute to an atmosphere in which it could be felt by ordinary people that they should hold views on such matters' (ibid.). The programmes in the single commissioned series contained two television presentations of contemporary debates in the history of science, *Portraits of Newton* (on changing representations of the natural philosopher) and *A History of Nature* (on the changing relationship of humans to the natural world). Other programmes were concerned with the radical politics of science, technology and medicine: *The Struggle for Health* (on health services in Zimbabwe); *I'm Not Ill, I'm Pregnant* (on the politics of pregnancy and obstetric technologies); *Stealing the Sun* (on the low priority accorded to research into renewable energy sources); and *Nuclear State* (on nuclear weapons).[3]

Following the publication of the Bodmer Report in 1985 and the establishment of the Committee on the Public Understanding of Science (COPUS) in 1986,[4] scientific organisations once again petitioned broadcasters about the televisual presentation of science, as they had between 1941 and 1962. Independently, Adam Curtis was developing a new series to extend political journalism into new areas, in this case science.[5] Therefore, after – but not because of – pressure from scientists, BBC2 showed *Pandora's Box* (1992). Thus the BBC had broadened responsibility for the production of science programmes beyond the successors to Aubrey Singer's Features and Science department of 1963, which had been the home for virtually all significant coverage of science. The subtitle of *Pandora's Box* was *Six Fables from the Age of Science*; this indicates the programmes' territory, which one critic describes as 'the consequences (often dangerous) of political and technocratic rationality' (Thomson 2005). The series consisted of six programmes: *The Engineers' Plot* (Soviet planning), *To the Brink of Eternity* (the Rand Corporation), *The League of Gentlemen* (the failures of economists' understanding of the British economy), *Goodbye, Mrs Ant* (DDT), *A is for Atom* (the rise and fall of nuclear power) and *Black Power* (Kwame Nkrumah's plan for an industrial utopia in Ghana centred on the Volta dam scheme). These programmes stand in very particular relation to the foregoing history of science films, in terms of their subject, the stress of their accounts and their cinematic technique. Some treated the themes that have occurred repeatedly in this account: planning; insects, hydro-electricity and nuclear power. However, their account of these subjects was vastly different from the techno-enthusiasm of the earlier films and television programmes. The programme on the Volta dam, for example, embodied a critique of these types of schemes that had been growing in intensity (see, for example, Rosenberg *et al.* 1995). Stylistically the *Pandora's Box* programmes also depart from most of what we have discussed in this book; these are authored documentaries, with Curtis's own voice providing the narration and interviewing witnesses. Its contrapuntal use of archive film displays the kind of cinematic literacy rarely seen since Rotha's big thematic films of the 1940s. Curtis's invasion of scientific television bears some comparison with Aubrey Singer's usurpation of Talks Department territory in the late 1950s, producing new genres of science television from outside the existing structure at a key moment. But Curtis, also like Rotha, was not only a scientific filmmaker and this goes some way to explaining the different cinematic literacy of these films and successors that also touch on scientific territory, such as *The Century of the Self* (2002).

In final conclusion

Cheese Mites and the rest of the Urban-Duncan oeuvre exemplify a mode of science filmmaking as a source of spectacle and pleasure. This mode, of rendering visible what science sees, has persisted for upwards of a century. Television coverage of cosmology is not dissimilar in its audience appeal and relation to science. The second mode, which celebrated the transformational power of science and technology, has also persisted, but the relationship of the public to this type of science has been indelibly affected by the impact of certain big technologies in the postwar technoscience era, especially nuclear power, nuclear weapons and agrochemicals. The result is that it is rare to witness science television that exhibits the techno-enthusiasm of a generation ago, except in ironic mode. Films in the tradition of *Enough to Eat?*, aiming to persuade audiences of all kinds that science is able to diagnose and cure major social problems, cannot avoid being affected by this public mood. Perhaps, with the current emergency of climate change, we stand on the brink of a new generation of films and programmes that will unironically and persuasively embody the voice of the scientist in this moral mode. Otherwise, television science, asserting its right to choose how it represents science 'as a process of television', may well have entered a period of crisis in which it oscillates between increasingly entertaining ways of conveying the content of science, reproducing the gaze of science (in natural history) or reporting scientific scandals. The age of the hortative pro-science film in the Rotha mode is long past. But, as I quoted Robert Young in the introduction, 'the future – including the very existence – of civilization depends on getting right the relations between expertise and democracy'. In Rotha's day, the appeal to citizenship was pervasive: make the case persuasive and democracy will vote for science and rationality. Now that such passion seems out of place, it is far from clear that viewers, pressing their red buttons, will opt for anything other than the latest equivalent of *Cheese Mites*. This is surely a burning issue for our 'advanced' scientific and technological society.

NOTES

Introduction

1 This is not the same meaning of the term 'genre' as employed more generally in film studies, for which see Grant 2006.
2 Rotha to Knight, April 1933, Paul Rotha Papers, Box 26 (Collection 2001). Department of Special Collections, Charles E. Young Research Library, University of California, Los Angeles.
3 See Boon (1993 and 1997) for more extended treatment.

Chapter 1

1 For the foundation of this account of Urban's films, I am indebted to Oliver Gaycken's excellent as yet unpublished thesis (2005). Readers are directed to this comprehensive work, not just for a closer reading, but also for a different interpretation of these films.
2 The copying of Doyen's film of the separation of Siamese twins for sale to fairground freak show exhibitors marks an only partial exception (see Christie 1994: 100).
3 Usage of the word 'scientist', coined by William Whewel in 1833 to 'collectively designate the students of the material world' slowly became accepted in the following decades. The new umbrella term 'science' (as opposed to 'the sciences') and the notion of a 'scientific community' were established by the end of the century, denoting a broad category of activity distinct from philosophy, theology, literature, and other aspects of learning with which they had previously been intermingled (Cahan 2003a; Morrell 1990: 983 *passim*).
4 This work builds on Richard Altick's pioneering *The Shows of London* (1978), which explored the culture of spectacle and provides a significant context for popular science.
5 In publications too, whether judgement is based on style of presentation, relative conciseness of popular accounts, theoretical commitments or nomenclature, distinctions are difficult to make (see Drouin & Bensaude-Vincent 1996: 409).

6 Presumably in the second edition of 1896. Duncan senior belonged to the cohort of nineteenth-century physicians who started as amateur scientists and became professionals. He specialised eventually in geology (see Gaycken 2005: 30).
7 John Pickstone has suggested three phases of 'ways of making' to complement his 'ways of knowing'; he speaks of craft, rationalised production and systematic invention (2000: 18–19).
8 Industrial interest films are discussed in the next chapter.
9 In France, by contrast, university-trained biologists including Jean Comandon and Jean Painlevé did make the films that were taken up and distributed (see Gaycken 2005; Landecker 2006).
10 See, for details, Screenonline: http://www.screenonline.org.uk/people/id/594315/index.html, where several films can also be seen by those in educational institutions and UK libraries.

Chapter 2

1 As I have argued, although in some measure these were 'about' disease, in this period they were both almost exclusively fictional in approach – I have called them 'moral tales' – and also very coy about disease processes (see Boon 1999 and 2005).
2 Martin Stollery's recent *Alternative Empires: European Modernist Cinemas and the Cultures of Imperialism* (2000) is an excellent exception to this general characteristic of the literature. Aspects of documentary treatment of industry as labour are treated in Dodd & Dodd (1996).
3 The Urban-Duncan films, for example, are not mentioned in Paul Rotha's primer *Documentary Film* (1936).
4 Most relevant amongst accounts of wider genres than just documentary are Rachael Low's works (1971, 1979a, 1979b); Bert Hogenkamp (1986, 2000), Stephen Jones (1987) and the contributors to *Traditions of Independence* (MacPherson 1980) have treated the films of the political left. The relations of politics and the newsreels have been the concern of a group of international historians at Leeds University (for example Nicholas Pronay, Phil Taylor). Scattered papers treat some other subject concerns – for example, Geoffrey Crothall's paper on slum clearance films (Crothall 1999).
5 A university associate of Grierson and member of the EMB and GPO film committees, Elliot was a familiar face at the documentarists' offices. As Minister of Agriculture (1932–36) and Health (1938–40), he was active in promoting the use of films (see Hardy 1979: 27, 49, 73; Rotha 1973: 43; Orr & Tallents 1958: 73–80).

6 The main biographical source is his autobiography (Rotha 1973), which is usefully supplemented both by Marris (1982) and by Petrie and Kruger (1999).
7 See Nicholls (1976) for a developed version of this analysis as 'modes of address'.
8 Rotha to Knight, 18 July 1934, Paul Rotha Papers, Box 26 (Collection 2001). Department of Special Collections, Charles E. Young Research Library, University of California, Los Angeles. Hereafter, 'Rotha to Knight'. Rotha modified the text when he quoted it in *Documentary Diary* (1973: 103).
9 It is generally accepted that documentarists were not alone in making this turn to concentrate on 'the social'. Photojournalism presented 'a new social reality: the domain of everyday life', and in Britain this found expression, from 1938 onwards, in *Picture Post* (Hall 1972: 83). In America, the photographic work of the Farm Security Administration displayed a comparable attention to the lives of the poor (see Tagg 1988: 168–75). The ethnographic and diary-keeping activities of Mass Observation have been seen in the same light (Chaney & Pickering 1986; Hall 1972: 98). A similar tendency is found in print journalism, such as Priestley's *English Journey* (1934) or Orwell's *Road to Wigan Pier* (1937), and in literature, for example Greenwood's *Love on the Dole* (1933) or Knight's *Now Pray We for Our Country* (1940).
10 Rotha to Knight, 7 October 1933.
11 Ibid., May 1933.
12 My understanding of these issues has been substantially enhanced by discussions with Ian Christie.
13 This timing is clear from his letters to Eric Knight. The commission for the book is mentioned in Rotha to Knight, 29 January 1934, and progress, for example in that of 8 August 1934, 3.
14 *Face of Britain* documents, Rotha Papers, Box 12.
15 This section draws on Boon (2000); readers are referred to this for a more extended reading of this film and its context.
16 There is a gap in Rotha's letters to Knight at this point, but the spring 1935 issue of *Cinema Quarterly* states that Rotha was 'at present directing *The Face of Britain*', whilst the summer issue includes a plate of stills from it, described as his new film, implying that it was completed in the spring.
17 He was a member of the committee of Political and Economic Planning that produced the 1936 report on electrical supply (see Hannah 1979: 250–1). He was also a member of the XYZ Club, which provided intelligence on city activities to the Labour Party and helped shape its

thinking on financial policy. Via that group, he was in contact with Hugh Gaitskell and Hugh Dalton (see Pimlott 1977: 37). (Gaitskell was a key figure in the development of arguments on behalf of economic planning at the New Fabian Research Bureau. See Durbin 1985, who devotes a chapter (8) to these issues.)

18 *Memorandum for the Production of a Programme of Documentary and Instructional Films on Architecture*, 1 February 1936, RIBA archives.
19 For a more detailed account, see Boon (2000).
20 Alternatively, the third and fourth sections may be seen together as the synthesis to the first two sections.
21 Rotha to Knight, 7 October 1933.
22 Sixth version of script, Rotha Papers, Box 12.
23 Ibid.
24 Fifth version of the script, Rotha Papers, Box 12.
25 Here the weight of 'scientific' is to denote 'rational' rather than to allude to the planning of scientific work itself, although that was a concern of some left scientists.
26 'The Face of Britain', two page summary with credits, Rotha Papers, Box 12.
27 It was conventional to see it as an event, rather than as a gradual process, the account favoured by more recent economic historians (see Cannadine 1984).
28 Contributors to Misa *et al.* (2003) assert the importance of a rapprochement between the macro-scale abstraction of modernity theory and the micro-scale empirical study of technologies typical of science and technology studies since the 1970s. This current study is not concerned with the relationship between technology and modernity in general, but with the ways in which people in the past constructed such a relationship. But Brey's analysis can help us; he has proposed two types of analysis to join the micro and the macro in terms of levels of abstraction. He describes as a *deductive analysis* one that interprets a phenomenon (in this case a film) as a species of a more general phenomenon (in this case, common components in 1930s modernist discourse). This is what my analysis of *The Face of Britain* achieves. He terms the analysis moving from the particular to the general a *specificatory analysis* (Brey 2003: 68).
29 The British Association for the Advancement of Science established a Division for Social and International Relations of Science (in the latter mode suggested by Elzinga), partially as a response to the rise of totalitarianism. See chapter three.
30 Or, alternatively, if one is willing to overlook the rhetorical hyperbole of Rotha's argument, it is not a documentary but an instructional film:

'to be truthful within the technical limits of the camera and microphone demands description, which is the aim of the instructional film, and not dramatisation, which is the qualification of the documentary method' (Rotha 1936a: 134).

31 See next chapter for details.

Chapter 3

1 Association was, for example, the dominant mode in health education (see Boon 1999).

2 Werskey's *The Visible College* (1988) and articles (particularly 1971) are concerned with individual scientists and the development of new hybrids of scientific thought and left political beliefs. A parallel account is Greta Jones' from the viewpoint of eugenically-inclined believers in social Darwinism and 'social hygiene' (1986). Others have looked at informal groupings in terms of their mixed membership and common concerns (for example, Zuckerman 1978). Peter Collins (1981) and William McGucken (1984) studied the organisational settings for debates on science and society, especially the British Association for the Advancement of Science (BA).

3 A phrase that Werskey found in Neil Wood's 1959 text, *Communism and British Intellectuals,* which he said 'seemed almost to dominate the British scientific world between 1932 and 1945' (1959: 121).

4 The name, produced in an exchange between J. B. S. Haldane and Lancelot Hogben – both members – derives from the Latin tag *tot homines quot sententiae,* which translates roughly as 'as many opinions as there are people'.

5 A film named *A Record of Progress* (c. 1936) went as far as being advertised, but was never released. It was described as 'a news reel review of the beginning of a new England. Smokeless cities, planned housing estates, more leisure for all ... the place of gas in national life' (Anon. 1936d: 195–6).

6 They made public relations an integral part of gas industry activity, not just through films, but also in public exhibitions, formal dinners with invited speakers and campaigns, for example on smoke pollution (see Boon 1993: 84–7).

7 The subsequent films dropped the claim of disinterest; *Plan for Living* (1938, discussed later) was made, as the credits stated, 'with the cooperation of Radiation Ltd', the gas cooker manufacturers, and *Daisy Bell Comes to Town* (1937), made with the National Milk Publicity Council, features not only a pantomime cow, but also a public cookery demonstra-

tion organised by the gas industry.

8 *Kine Weekly* commented, for example that 'it is entirely free of sordid and depressing scenes, but shows the public efforts in feeding British children and their mothers, with very telling and sensible expositions' (Anon. 1936i: 23). The *Times* opined that 'while the lesson of the film is vigorous and direct, the producers must be complimented on their avoidance of extreme cases' (Anon.1936j: 8).

9 A letter from Edgar Anstey to Julian Huxley confirms that he received a fee for narrating *Enough to Eat?*; Anstey to Huxley, 26 October 1936, Julian Sorell Huxley – Papers, 1899–1980, MS 50, Woodson Research Center, Fondren Library, Rice University (hereafter 'Huxley Archive'), Box 12:3.

10 Leaflet preserved with letter Edgar Anstey to Edward Mellanby, 24 June 1936, National Archives (NA) FD 1/3427. For ARFP see Rotha (1939) and Anon. 1936e. The date 'late 1935' is given in *Arts Enquiry* 1947: 55. The first suggestion seems to have come from Grierson to Rotha, but several of the individuals involved were Rotha's contacts, and the regular business of the group seems to have been conducted by him (see Rotha 1973: 160–1).

11 I have found no documents that explain how Huxley, Haldane and Hogben became ARFP advisors, but it is plausible that Ritchie Calder was the link.

12 Other figures significant to our story, including Richard Gregory (editor of the scientific journal *Nature* and key figure in the British Association) were also involved (Jones 1986: 106, 111).

13 The ARFP minute book was lost (Rotha 1973: 161).

14 Some of Huxley's appointment diaries have entries for Film Society showings; Huxley Archive, Box 49.

15 See Watson Watt & Keen (1936); J. Huxley, 'British Films (letter to the editor)', *The Times*, 22 June 1937: 17.

16 Hogben to Huxley, 20 Oct. 1936, Huxley Archive.

17 His personal archive is eloquent about his tireless activity; his correspondence, publications and press cuttings reveal his wide activities and interests. The most eloquent published account is the collection edited by Waters and Van Helden (1992).

18 The report he wrote on students' responses led to the more formal Bantu educational film experiment (Low 1979a: 43-4; Burns 2000: 202–3).

19 Conference on Educational Films, 12 April 1929, NA CO758/91/9.

20 The obvious comparisons are *Ravenous Roger* (1935), on the life cycle of the raven and *Secrets of Nature: The Gannet* (1923), filmed by Oliver Pike.

21 ARFP *Syllabus of Lectures on the Film Prepared and Issued by Associated Realist Film Producers*, London, 1937–38. Rotha Papers, Box 80. See also Anon. 1937f: 8.
22 Report of Meeting, 3 October 1935, SA/EUG D.71. Contemporary Medical Archives Centre, Wellcome Library, London.
23 C. P. Blacker to Huxley, 23 February 1937, SA/EUG D.74.
24 3 October 1935, when a meeting between Huxley, H. R. Hewer (of Imperial College and scientific films adviser to G-B I) and C. P. Blacker, the Society's General Secretary, discussed a fairly fully formed structure. Report of Meeting, 3 October 1935, SA/EUG D.71.
25 See *Monthly Film Bulletin* 1939: 6, 28, and G-B I Booklet accompanying film, in SA/EUG.D71 & D.72.
26 G-B I Booklet accompanying film, in SA/EUG.D72.
27 Report by HFP 9 July 1946 on propaganda films from 1937, SA/EUG. D72.
28 Undated note headed 'Film, "Heredity in Man"', SA/EUG.D74.
29 Estimates of the health of the British Eugenics Society in the 1930s vary. However, it is clear that C. P. Blacker, the General Secretary, had a difficult job sustaining its political balance, especially retaining its predominantly conservative membership, in decline from the middle of the decade, whilst seeking to retain the support of the scientific community, especially of those such as Haldane, Hogben and Huxley who were professionally concerned with biology and genetics (Jones 1986: 88, 108). By the mid-1930s, however, the trend in biology was away from the type of simple hereditarian arguments used by the Eugenics Society towards a more environmental stress (see Searle 1979: 163–6).
30 Précis of film dated 8 June 1937, embodying JH's corrections, SA/EUG. D71.
31 Blacker to Hewer, 5 May 1936 (see also Hewer to HFP 9 April 1936, SA/EUG.D71); '…As you suggest, the case bristles with difficulties not only in the legal way but also in the biological…'
32 For example, the Chief Medical Officer George Newman wrote to the Minister of Health in January 1934 that a joint meeting between the two committees should not be held because 'it will involve the Ministry in a far-reaching economic issue, which is most important to avoid – an issue which might easily affect wages, cost of food, doles, etc' (quoted in Mayhew 1988: 451).
33 When Orr wanted to add a meeting room to the Rowett, Elliot 'invited Orr to lunch in London with Lord Strathcona, a man with great interest in Dominion relationships and also blessed with considerable means. As a result, Strathcona offered £5,000, later increased to £8,000 … Wal-

ter Elliot … provided carpets for the public rooms' (Kay 1972: 51, 54). David Smith has shown that Orr slowly broadened the concerns of the Rowett from an initial focus on animal nutrition to take in human nutrition and its relation to agricultural economics and marketing (2000: 64).

34 Also a member of the Next Five Years Group, which Marwick identified as a 'Middle Opinion' organisation (see Jones 1986: 113–33).

35 Annex to PEP Bulletin 4, 30 December 1933, PEP/A/6, PEP Papers, LSE Archives.

36 Second draft *Memorandum on the Need for a National Food Policy*, 10 December 1934, PEP/WG/2/1, PEP Papers.

37 It continues: 'where doubts are still felt about the need for planning, the approach to a national food policy … should prove significant … The magnitude and suddenness of the adjustments needed in this field exceed anything with which an unplanned system has ever had to cope…'

38 9 October 1934. Transcript in PEP/WG/2/1, PEP Papers.

39 See PEP Research Group minutes of meeting, 5 November 1934 in NA file FD 1/3532, 'Political and Economic Planning (PEP) Group: memorandum on need for a national food policy, 1934'.

40 For example, Edward Mellanby (since 1933 Secretary of the Medical Research Council) had argued that 'proper feeding … would be as revolutionary in its effects on public health and physique as was the introduction of cleanliness and drainage in the last century' (quoted in MacNicol 1980: 46). Others took an unreductionist holistic approach; the 'clinical art' (see Lawrence 1998).

41 This is a description of the extant copies of the film. According to *Monthly Film Bulletin*, there was originally a 'longer version' at 2,500 ft (three reels) as well as the shorter at 2,008 ft (two reels) (see Vesselo 1936).

42 Anstey to Mellanby, 24 June 1936, NA FD 1/3427.

43 It is possible that the clumsiness visible at this point betrays a cut between the longer and this shorter version of the film.

44 Pamphlet says 'country'; film says 'world'.

45 'Nutrition Film (5th Draft)' attached to Anstey to Mellanby, 24 June 1936, 7, NA FD 1/3427.

46 Anstey interviewed in 'Extra! New British Edition', *On the March* (series on the history of the *March of Time* Newsreel), Flashbacks Productions (30 October 1985).

47 Anstey to Mellanby, 24 June 1936, NA FD 1/3427. This emphasis on the scientific nature of the film was driven home in the published version of the script, which lists nine source books and pamphlets, including books

by Orr, M'Gonigle, Friend and McCarrison as well as a pamphlet by the Committee Against Malnutrition.

48 The left journalist Claude Cockburn lampooned Astor and his Tory MP wife Nancy as leaders of 'the Cliveden Set', a group of wealthy Conservative power-brokers who met at their country house in the 1930s, and were seen as having a powerful influence over government (see Leventhal 1995: 39, 163).

49 W.A.R. to Minister, 11 January 1934, NA MH 56/56.

50 It probably came about via his association with Elliot, a key member both of the Board's Research and Film committees. Elliot was chairman of the film committee in 1927. The pasture film was announced to this body on 23 July 1929. See NA CO 760/37. This film was a particularly striking instance of a project bridging the scientific and public relations functions of the Board. The nutritional qualities of grazing grasses were extensively discussed, as for example at a meeting in October 1927, where Elliot, Orr and Huxley were all present (Discussion Held on 6 October 1927, Concerning the Empire's Grassland Problems, NA CO 758/41/1.) It was reported in July 1929 that 'Mr Grierson had already discussed the suggested theme of the film with Dr Orr … who was very helpful as to the possibilities and utility of such a film'. It was cancelled within six months because 'it had been ascertained that certain elements had not yet been demonstrated on a commercial scale'. (Reported at film committee meeting, 30 January 1930. Several draft scripts survive, ibid., and Grierson Archive, Stirling University, G2.2).

51 Re-released in 1933 with a soundtrack under the title *Shadow on the Mountain*, directed by Arthur Elton and showing the work of the Welsh Plant Breeding Station at Aberystwyth under Professor George Stapledon in the breeding of nutritious grazing grasses (Low 1979a: 55–6).

52 Anstey to Mellanby, 27 July 1936, NA FD 1/3427.

53 Mellanby to Anstey, 6 August 1936, NA FD 1/3427.

54 'Nutrition Film (5th Draft)' attached to Anstey to Mellanby, 24 June 1936: 6. NA FD 1/3427.

55 K. McGregor to H. Leggett, 27 August. 1936, MH 78/236.

56 Anstey to Mellanby: 4–5, 24 June 1936, NA FD 1/3427.

57 Huxley journal 8 October 1936, Huxley Archive, Box 49:5.

58 Anstey to Huxley, 20 October 1936, Huxley Archive, Box 12:3.

59 In the significantly different American version of the story, distributed in April 1937, under the title *Britain's Food Defenses*, Kingsley Wood is seen. In a bridge between Johnston (identified as a socialist) and Wood, the commentary states 'as the opposition and radical press take up the cause, the Conservative party's Sir Kingsley Wood, Minister of Health,

accuses the Socialists of distorting facts, says that the problem is not too little food, but the wrong food'. This can be seen online at http://xroads.virginia.edu/~MA04/wood/mot/html/timeline.htm.

60 A similar argument might be made about *Pots and Plans* (1937), which applied motion study to the design of a kitchen.

61 For a more extended treatment see Boon (1993).

62 In the Muirhead Lectures on political philosophy delivered in the University of Birmingham in January and February 1938, he commented 'I have only been a Marxist for about a year' (Haldane 1938: 13).

63 This included several more of the most prominent architectural modernists active in Britain, including Wells Coates, Berthold Lubetkin and Francis Yorke (Darling 2000: 165–71).

Chapter 4

1 It was completed as far as show-print stage by May 1940; Rotha to Grierson, 30 May 1940; Grierson Archive, University of Stirling, 4:24:133. This film was not released at the time and was not shown until 1965 (Rotha 1973: 263).

2 Rotha to Grierson, 27 August 1940; Grierson Archive, 4:24:154.

3 Rotha, 'Report on the Production and Distribution of Documentary Films for the Purpose of Propaganda and Information by the Ministry of Information, Part II, Film Production Programme For Great Britain': 5–6, Paul Rotha Papers, Box 69 (Collection 2001). Department of Special Collections, Charles E. Young Research Library, University of California, Los Angeles.

4 Rotha's diary for 6 September confirms him to be the author; Rotha Papers, Box 82 (hereafter, Rotha Diary).

5 Huxley gave a first report at the BA annual conference in Dundee as war broke out, where once again he stressed how unplanned science was, and how unconnected to the conditions of its consumption (see Anon. 1939b).

6 The documentarists' organisation that had assumed the contact brokering duties of ARFP in 1937 (see Swann 1989: 76).

7 There is a copy of this script in the Rotha Papers, Box 1.

8 Huxley, 'Planning of London: Memorandum of interview with Sir Stephen Tallents and Mr Edward Carter 11 October 1940'. London School of Economics PEP papers WG19/1.

9 This conference suggested the topics of two further meetings: in March 1942 on the subject of postwar agriculture and in July on mineral resources in the light of the Atlantic charter.

10 Williams edited in 1941 'The Democratic Order', a book series which included volumes by Ritchie Calder and Huxley (Ritchie Calder 1941; Huxley 1941).

11 Peter Ritchie Calder Diary, 3 February 1940, National Library of Scotland, Edinburgh, Manuscript Division, Dep. 370/186.

12 Other members included Bernal (see Brown 2005: 283), architect and planner Patrick Abercrombie, town planner Frederick Osborn, L. V. Easterbrook (Agriculture correspondent of the *News Chronicle*), housing expert Elizabeth Denby (the planner behind Kensal House), Major Sandford Carter (probably from the National Council of Social Service), George Pepler and Ronald Davison, with A. M. Carr Saunders (of PEP and the Eugenics Society) and Orr invited to attend. Extra detail from Huxley's note of discussion with Carter and Tallents, PEP WG/10/1.

13 E. J. Carter, *The Limits of the 1940 Council's Mandate*, covered by note to Sir Ian MacAlister, 30 August 1940, typescript memorandum, EJ CaE/2/2, Memorandum of interview with Detective Inspector Smith, 22 July 1940, CaE/2/3, EJ Carter papers, RIBA Archive.

14 E. J. Carter, *The Limits of the 1940 Council's Mandate*, 30 August 1940, typescript memorandum. Following this stress on publicity, one of the most visible of the 1940 Council's activities was the *Living in Cities* exhibition, produced collaboratively with the British Institute of Adult Education in May 1941. Curated by Council member and architect Ralph Tubbs, later designer of the Festival of Britain's 'Dome of Discovery', it was shown at RIBA, with three copies circulated on a tour of public libraries; Tubbs 1942; *Architectural Review* 1941: 143–4; Rotha Diary 7 May 1941.

15 Rotha Diary, 3 October 1940.

16 Ibid., 4 October 1940.

17 These included Zuckerman, Bernal, Carter, Crowther, Huxley, C. D. Darlington (then Director of the John Innes Horticultural Institute) and C. H. Waddington (see Zuckerman 1978: 111–12).

18 Rotha Diary, 8 October 1940. All related quotes on this page are from this source.

19 Ibid., 22 December 1940; 8 January 1941.

20 Ibid., 8 January 1941.

21 Rotha, *Documentary Films Yesterday and Today*, press release appended to P. Rotha to J. Huxley, 11 February 1941, Julian Sorell Huxley – Papers, 1899–1980, MS 50, Woodson Research Center, Fondren Library, Rice University (hereafter 'Huxley Archive').

22 Rotha was not alone in believing that science films would benefit wartime propaganda; in parallel with the start-up of his company, Sidney

Bernstein from the MoI had requested from Huxley a list of scientific subjects for films. Huxley's response retained the emphasis of the Rockefeller initiative on diet and nutrition from the year before and included two subjects that Rotha later included in films: blood transfusion and diet in wartime. The full shortlist considered for production was: 'blood transfusion, diet in wartime, food preservation, new sources of energy, hormones, training of airmen, plastics, industrial psychology, genetics of race, new methods of wound treating, textile research and meteorology'; S. Bernstein to J. Huxley, 7 February 1941 (see also Huxley to Rotha, 13 February 1941), Zuckerman Archive, University of East Anglia, SZ/TQ/251 (hereafter 'SZ/' citation).

23 Rotha Diary, 10 February 1941.

24 Minutes of Tots and Quots meeting, 27 February 1941, SZ/TQ/1/7.

25 Rotha Diary, 2/3 April 1941.

26 Its members were Ritchie Calder, Carter, Easterbrook, Huxley, David Owen (PEP's General Secretary and soon to become Stafford Cripps' personal assistant), Basil Ward, W. E. Williams (Army Bureau of Current Affairs and Institute of Adult Education) and the documentarist Basil Wright; Rotha Diary, 11 March 1941.

27 Rotha Diary, 4 April 1941.

28 Rotha to Grierson, 19 March 1941, Grierson Archive, 4:25:6.

29 Rotha Diary, 17 February, 1941.

30 Ibid., 19 June 1941.

31 Ibid., 22 June 1941.

32 Rotha to Knight, 19 July 1941, Rotha Papers, Box 26.

33 Rotha Diary, 24 July 1941.

34 Ibid., 17 August 1941. See also Rotha, *Chairman's Report to Shareholders*, 3 May 1942, Rotha Papers, Box 72.

35 Rotha to Zuckerman, 7 August 1941, SZ/TQ/2/3/3.

36 Rotha to Carter, 4 August 1941, E.J. Carter papers, Ca E/2/3. Rotha to Zuckerman, 7 August 1941, SZ/TQ/2/3/3; Zuckerman to Rotha 12 August 1941, SZ/TQ/2/3/4. There are copies of the script in both archives.

37 Paul Rotha Productions Ltd, *First Treatment for a Non-Theatrical Film on 'Science and War'*, Preserved in E. J. Carter papers, RIBA Archives.

38 This was promoted as the transformation of winter wheat into spring wheat by the exposure of germinating grains to cold. This was subsequently discredited.

39 Amongst possible explanations is the fact that Michael Orrom, who had been developing the script, was conscripted to the radar corps in August 1941 at a time when Rotha's company was very busy (Huxley to

Rotha, 6 August and 16 August 1941, Huxley Archive). Ritchie Calder cannot have been able to give as much to the project as they might have wished, as early in 1941 he had been recruited to work for the Political Warfare Executive, a branch of the Government's Special Operations Executive devoted to providing counter propaganda in Germany and Axis-occupied countries, Rotha Diary, 17/18 August 1941; *The Times*, 2 February 1982, 12. His attitudes towards the communication of science to the public cannot but have been altered by this secret appointment. Rotha's speech to his first annual meeting, which covered this period, commented that 'we depended largely on contracts from the Ministry of Information, which body was still in a state of sorting itself out. Much time was wasted on films allocated to us from this body which were later cancelled' (*Chairman's Report to Shareholders*, Rotha Papers, Box 72). We must assume that *Science and War* was one such.

40 Rotha Diary, 17 August 1941.

41 Neurath had already used Isotypes at the Museum of Economy and Society in Vienna. After he settled in Britain, he became interested in the active engagement of citizens in decision-making about planning, resonating with, and then extending, the citizenship ideology of his British collaborators, including Rotha (see Nikolow 2004).

42 Rotha to Neurath, 7 April 1941, Isotype Collection, University of Reading Manuscript collection.

43 Neurath to Rotha, 9 April 1941, Isotype Collection.

44 Rotha Diary, 16 May 1941; 7 May 1941.

45 See, for example, Neurath to Rotha, 5 November 1942, Rotha Papers, Box 74.

46 See for example, of many references, Neurath to Rotha, 25 September 1942.

47 Second draft script, March 1942, NA INF 1/214 pt.1: 137.

48 Rotha to Pudovkin, 28 March 1945, Rotha Papers, Box 66.

49 Rotha to Knight, 6 May 1942.

50 Rotha to Pudovkin, 28 March 1945, Rotha Papers, Box 66.

51 Rotha to Elton, 15 October 1941, NA INF 1/214 pt. 1: 32. Rotha cited *The Strategy of Metals*, a film made by Raymond Spottiswoode and Stuart Legg for the National Film Board of Canada in 1941, which was concerned with the strategic importance of Canadian aluminium as a war resource for the allies.

52 Marshall suggested two versions on 2 February 1942, NA INF 1/214 pt. 1: 85. Priority was given to the American version and the idea of a British version was dropped only after the American version was complete.

53 Preserved at NA INF 1/214 pt. 1: 45–69.

54 For Knight's letters see Knight 1952. Both sides of the correspondence are preserved in the Rotha Papers.
55 Earlier, for example, with Lancelot Hogben, H. G. Wells and others, Marshall had been on the editorial panel of Fact, a sort of left Book of the Month club (see 1937a). For obituaries of Marshall see Anon. 1992a; Anon. 1992b).
56 Marshall to Elton, 14 January 1942, NA INF 1/214 pt. 1: 31.
57 Rotha to Neurath, 5, 11 February; 3, 31 March 1942.
58 Rotha to Marshall, 2 February 1942; Marshall to Beddington, 20 March 1942; Duff to Campbell, 8 April 1942, NA INF 1/214 pt. 1: 100, 169.
59 Rotha Diary, 10 May 1942.
60 Rotha to Marshall, 27 October 1942, NA INF 1/214 pt. 2: 226–8.
61 Rotha to J. Griggs, Ministry of Information, 12 January 1943, NA INF 1/214 pt. 3: 41.
62 Rotha Diary, 17 May 1942. Rotha to Pudovkin, 28 March 1945, Rotha Papers, Box 66.
63 Rotha also sent to him a memorandum on postwar documentary at this time; Rotha Diary, 13 July 1942. See chapter five.
64 The disputes discussed here briefly have a more detailed coverage in Boon 1997.
65 Elton to Marshall, 29 October 1942; see also Rotha to Marshall, 24 November 1942, NA INF 1/214 pt. 2: 229–30, 233.
66 Elton to Marshall, 29 October 1942, INF 1/214 pt 2: 229.
67 Bernstein to Beddington, 26 August, NA INF 1/214 pt. 2: 168–70.
68 Ibid.; Beddington cable to Bernstein, 31 August; Rotha to Beddington, 7 September 1942, NA INF 1/214 pt. 2, 192, 199–200. Rotha also agreed to excise a section of part two featuring the poor physical condition of American army recruits.
69 Calder-Marshall to Beddington, 4 February 1942, NA INF 1/214 pt. 1: 85.
70 Orr's account of Woolton's reported reason for not having set up a panel of experts to monitor food supplies implies such a conflict: 'Lord Woolton then said, "I had it in mind, but it was opposed by the Minister of Agriculture"' (Orr 1966: 120).
71 Westerby to Rotha Productions, 13 June; Rotha to Horder, 27 May 1942, NA INF 1/214 pt. 2: 71.
72 Manktelow to Calder-Marshall, 11 April 1942, NA INF 1/214 pt. 2: 130.
73 S. Smith, Director of Public Relations, Ministry of Food, to Beddington, 7 May; Elton to Beddington, 12 May 1942, NA INF 1/214 pt. 3: 170, 167, 173.
74 Elton to Mercier, 7 May 1943, INF 1/214 pt. 3: 175.

75 The published script follows the excised version but both National Film Archive copies that I have seen retain these sequences.
76 Elton to Beddington, 12 May. All following quotes are from this source.
77 Gates to Beddington, 13 May; see also Vandepeer to Beddington, 24 May 1943, NA INF 1/214 pt. 3.
78 Judged on the basis of the list printed in Marris 1982, 96–109.
79 Rotha to Knight, 6 July 1942.
80 27 February 1941, production memo in file INF6/517; 'Diphtheria Immunisation Campaign' NA ROA 404, MH 101/30.
81 Social scientist and director of home publicity at the MoI, who would have been familiar to viewers from his popular radio broadcasts and column in the *News Chronicle*; see Hilton 2006.
82 Script preserved in INF6/517.
83 Memo by JKR, 25 September 1951, INF6/520
84 Shooting script 3.2/266–268 preserved in Neurath/Isotype collection (MS109), University of Reading Archives. Final script in NA INF 6/1934.
85 This film was also in many ways the expression of Rotha's long-nurtured wish to make a film on themes of urban planning, architecture and housing. He had worked on a script, *The Growth of a City*, a film on architecture and town planning, immediately after reading Lewis Mumford's *The Culture of Cities* soon after it came out in 1937. He wrote to Knight at the time that it 'will tell how Everytown, 1938, England, grew up; how housing and public services have been related, or in so many cases unrelated, to the needs of the community. It could be something for which *The Face of Britain* was a child's notebook'; Rotha to Knight, 20 December 1938. After many changes, these ideas surfaced first in John Taylor's film *Goodbye Yesterday* (1941), which was completed but suppressed by the MoI, and finally here (Rotha 1973: 228–9).
86 Rotha Diary, 7 May 1941.
87 Ibid., 18 August 1941.
88 Ibid.
89 Ibid., 10 May; 13 July 1942.
90 Rotha to Knight, 6 July 1942, 5.
91 Rotha Diary, 13 July 1942.
92 Ibid., 25 July 1942.
93 It is referred to as 'this world planning script' in Rotha Diary, 17 March 1942.
94 Rotha Diary, 18 October 1943; Rotha to Neurath, 8 October 1943.
95 Ibid., 25 November 1943.
96 Ibid., 3 September, 17 December 1943; Rotha to Neurath, 8 June 1944,

10 February 1945.

97 Although Rotha had to subsidise its completion from his company (see Pearson 1982: 68).

98 At Neurath's request they investigated using an electronically-treated voice; Neurath to Rotha, 2 December 1942, Rotha Papers, Box 74.

Chapter 5

1 Rotha Diary, 13 July 1942, Paul Rotha Papers, Box 23 (Collection 2001). Department of Special Collections, Charles E. Young Research Library, University of California, Los Angeles. Something similar later briefly came into existence as 'Group 3', where Grierson was employed for a while (Ellis 2000: 274–90).

2 The completed films are only a proportion of those developed or contemplated in this period. In 1945, he discussed with Neurath a film on the atom bomb (Neurath to Rotha, 14 August 1945, Rotha Papers, Box 74). Examples of abandoned government films include the subjects of building sites, the new agriculture, nursing and young scientists; 'Film Projects Abandoned at the Scripting Stage', National Archives (hereafter NA) INF 12/296; 'Contracts Placed with Films of Fact Ltd', INF 12/122. But the reputation of his films also led others to seek to employ him for particular projects. For instance, in 1947, he, along with Basil Wright, was favoured by the Ministry of Health to make a substantial four-reel film to promote the new National Health Service (see Wildy 1986: 13).

3 John Wales to Robert Fraser, 25 September 1947, INF 12/122.

4 Memorandum, *Britain Can Make It*, INF 6/592.

5 Ibid.

6 A title now most often associated with the 1947 Victoria and Albert Museum exhibition of British design, promoted by Cripps's Board of Trade, also the subject of issue 12 of this newsreel.

7 See the review in *Monthly Film Bulletin* (Anon. 1946c: 11). See also the review in *Documentary News Letter* (Anon. 1946d: 24).

8 A project in which J. D. Bernal was involved.

9 Cripps had become an apostle of motion study when Minister of Aircraft Production (MAP) from late 1942, where he had formed a Production Efficiency Board (PEB) to which Anne Shaw – who had conducted this kind of work in the electrical industry in the inter-war period – was attached, with motion study one of the main approaches employed. When Cripps went to the Board of Trade after the 1945 election he took the PEB with him and changed its name to the Production Efficiency Ser-

vice, making it a management consultancy aimed mainly at small private firms. Jim Tomlinson (1992) describes the Service as a key instrument of postwar Labour Party industrial policy.

10 For example, 'Post War Organisation of Government Publicity', 18 September 1945; see NA T 222/66.

11 *The World is Rich,* file note by RM, 31 July 1950, INF 6/728.

12 'Films Abandoned at the Scripting Stage During 1947/8', 26 October 1948, NA INF12/296.

13 These included *Today and Tomorrow* (1945) about the wartime organisation of food supplies in the Middle East and, in the same year, *The Star and The Sand* about the 30,000 Yugoslav refugees evacuated to UNRRA camps in Egypt. See Thorpe and Pronay (1980). Earlier, he had written *White Battle Front* (1940) about the Army medical services.

14 Rotha to De Mouilpied, 29 May 1946, Rotha Papers, Box 12. Orr had shown his commitment to Rotha in 1944 by becoming a director of Films of Fact (with Richie Calder and H. E. Beales of the London School of Economics).

15 Orrom's unpublished autobiography (kept in the archive department of the BFI Library) alludes to a senior meeting on 18 March.

16 The sequence is described in 'Note on the various stages in the production of a Central Office of Information Film', Appendix E to B. C. Sendall, 'Relations between the Documentary Film Industry and the Central Office of Information', October 1947, INF 12/564.

17 J. Wales to A. G. Anderson, 4 July 1946, INF 6/728.

18 C. L. Paine to J Wales, 25 February 1946, ibid.

19 Rotha to De Mouilpied, 29 May 1946, Rotha Papers, Box 12.

20 Philip Mackie (COI) to Rotha, 16 July 1946, ibid.

21 Rotha to De Mouilpied, 6 January 1947, ibid.

22 Films of Fact to De Mouilpied, 15 April 1947, ibid.

23 Tritton to Sendall, 25 July 1947, NA INF 12/564.

24 Although we may note that the drama documentary *Children on Trial* (1946) cost £24,800 (see Robinson 1957a: 8).

25 B. C. Sendall to Rotha, 26 November 1946; Tritton to Sendall, 8 November, INF 12/122.

26 Tritton to Rotha, 23 May 1947, Rotha Papers, Box 73, *World is Rich* letters.

27 H. De Mouilpied to Rotha, 23 May 1947, ibid.

28 Rotha to Tritton and De Mouilpied, 28 May 1947, ibid.

29 Ibid.

30 First treatment for the film, 27 May 1947, 4, Rotha Papers, Box 12.

31 Ibid.

32 Ibid.
33 Rotha has corrected Calder-Marshall's typescript in the Rotha Papers copy.
34 Reisz and Millar (1972: 203–9) give a detailed exposition of the montage structure of this prologue.
35 Rotha to Tritton, 18 October 1947, Rotha Papers, Box 73, *The World is Rich* letters.
36 Gove Hambidge (Director of Information at FAO) to Rotha, 3 September 1947, Rotha Papers, Box 73.
37 Rotha to Tritton, 18 October 1947, Rotha Papers, Box 73.
38 Tritton to Rotha, 16 October 1947, ibid.
39 Ibid., 20 Oct 1947, ibid.
40 Rotha to Wyatt, 22 October; Wyatt to Rotha, 30 October 1947, ibid.
41 Editorial, *Daily Herald*, 14 Feb, 1948; De Mouilpied to Rotha, 10 December 1947, Tritton to Rotha, 20 January 1948, Rotha to Tritton, 22 January 1948, ibid.
42 Hansard, D. C. Deb 5. S, vol. 450 Oral Answers, 4 May 1948, 1097.
43 Description from translation of review in *Basler Nachrichten*, Rotha Archive Box 12/73.
44 'Overseas Planning Committee: Summary of World Common Themes', 17 January 1946, NA CAB 134/544. See Taylor 1989 for a discussion of the context of this policy. Another of these films, a similarly curious choice of subject, was *One Man's Story* (1948), a small-scale biopic of George M'Gonigle, the Medical Officer of Health for Stockton, who had caused the Ministry of Health such embarrassment over the subject of affordable diets in the mid-1930s, as we saw in chapter three.
45 Issues 11 and 14; BFI Film and Television database.
46 The details of this phase are in NA INF 12/123.
47 Paine to Ferguson, 30 September 1946, INF 12/122.
48 Memoranda throughout INF 12/122.
49 Robert Fraser, memo to Woodburn, 13 July 1947, INF 12/122.
50 Tritton to Sendall, 1 July 1947, INF 12/564.
51 De Mouilpied and Forman's paper, INF 12/564. This reference to the BBC's new cultural radio service, later renamed 'Radio 3', related to contemporary debates about élite and mass audiences (this radio station did not, at the start, concentrate exclusively on classical music; see Carpenter 1997).
52 Sendall to DG, 28 August 1947, INF 12/564.
53 Ivor Montagu's *Man – One Family* (1946) is one exception, as it has 'an ordinary European citizen' interrogating the commentator twice in a film which, with animated diagrams and stock footage, looks like a Rotha

copy.

54 Appendix A to: B. C. Sendall, 'Relations between the Documentary Film Industry and the Central Office of Information', October 1947, INF 12/564.
55 Tritton to Sendall, 1 July 1947, INF 12/564.
56 Tritton, *Background to the Documentary Situation*, 25 July 1947, INF 12/564.
57 Huxley himself continued a small output in films and television, including *Man – One Family* (1946) and participating in the BBC's 1954 BA programme, *Parliament of Science.*
58 He named the middle film *The Silent Revolution*, but it is *The Peaceful Revolution* that fits the description.
59 For example, in the NCB's *Mining Review* newsreel, we encounter the parents of a Coal Board apprentice discussing their son's aptitude for the job (*Mining Review*, 19, 3, November 1965).

Chapter 6

1 For example, in articles on science broadcasting for the *BBC Quarterly* in 1949–50 neither the nutritionist V. H. Mottram nor Ritchie Calder mentioned television, even in terms of its future potential.
2 An early example was Gerald Heard's radio series *This Surprising World* (1930–32). He preferred to take advice *ad hoc* from Daniel Hall and Richard Gregory of the nominal seven scientists; Anon., 'Memo: Advisory Panel on Scientific Subjects', undated, BBC Written Archives R6/288. (All archival references in this chapter are from here unless otherwise specified.) Ralph Desmarais (2004) tabulates all BBC science radio programmes in his period.
3 The fact that Huxley addressed Ogilvie as 'Freddie' gives an indication of how well-connected Huxley was.
4 'Science and the BBC', Memo sent by Reinet Fremlin (secretary of the AScW) to O. J. R. Howarth (BA) and forwarded by Huxley to Ogilvie at the BBC, late October/early November 1941; R51/397/6.
5 Undated memorandum, R51/529.
6 Ogilvie cited a similar arrangement with the British Medical Association for medical stories, R51/529.
7 See also Desmarais 2004: 66–8.
8 R. Gregory *et al.* to Director General, BBC, 19 November 1943, R51/529
9 Basil Nicholls to A/C Ent, 15 August 1945, R6/288/1. There were further initiatives in 1944. On 26 October 1944, Sir Ernest Graham Little,

physician and MP for the University of London asked a Parliamentary Question on the matter. The attitude of the BBC hierarchy was the same when similar initiatives came from within. For example, Basil Nicholls, Senior Controller of Home Output, dismissed a request to establish an advisory committee for radio science features and a science magazine programme; Basil Nicholls to Acting Controller, Entertainment, 20 August 1945, R6/288/1.

10 The BBC did have a body they called a Scientific Advisory Committee from 1950 onwards, which advised on technical and scientific aspects of broadcasting technique, not on the representation of science in programmes.

11 M. L. Oliphant, 'The Broadcasting of Science', 16 May 1949, R6/34.

12 "The Presentation of Science', Report on a discussion initiated by Professor Oliphant', 2 June 1949, R6/34. His reference to the evils of science was in all likelihood about nuclear weapons. His wish to 'forget war and atomic weapons' is congruent with this; politics should not, in his view, contaminate the broadcast coverage of science.

13 Ibid.

14 'BBC General Advisory Council Report of Special Sub-Committee to consider Broadcasts on Science', 23 November 1949: 1, R6/34.

15 This reported back to the General Advisory Council at the end of November 1949.

16 They silently dropped the request for an advisory committee and for more scientifically qualified producers. Ibid., 2–3.

17 Barnes (Director, Spoken Word – a post bridging radio and television) to Oliphant, Hill and Bragg, 22 December 1949, R6/186.

18 A. V. Hill to Barnes, 26 December 1949, R6/186.

19 H. Dale, Report to the BBC Governors, 13 January 1953: 8, R6/186.

20 Ibid.: 9.

21 Since the war years, the Corporation had seen a distinction, seeking advice on medical – as separate from scientific – matters from the British Medical Association; see note six.

22 H. Dale, Report to the BBC Governors, 13 January 1953: 10, R6/186.

23 Mary Somerville (Controller, Talks (Home Sound)) to DSW 'HD's Report', 26 January 1953, R6/186.

24 Mary Somerville to DSW 'Sir Henry Dale: Advice on Scientific Programmes', 9 September 1954, R6/186.

25 The BBC's technical historian shows that they invested heavily in the technology so that they had 53 machines by 1972 (see Pawley 1972: 494).

26 Also Deputy Director of BBC Television until 1961.

27 This is discussed in the introduction to Rotha 1956.

28 'Science and the radio, a conference at Cambridge, 18–19 May 1946. C. Science in television programmes, Mary Adams'; R51/529.
29 Ibid. She lists some specific examples.
30 A. Powell, 'Opening Address', 20 March 1943, R51/529.
31 Robert Barr to Controller, Television programmes, 'T/V Documentary', 3 August 1951, T16/61/1.
32 McCloy to McGivern, 'Presentation of Science on Television', 7 October 1958, T16/623.
33 *Passim*, T32/216.
34 See Pound 1958: 724.
35 G. Boumphrey, L. Hardern, memorandum 'Inventors' Club', March 1953, T32/216/4. For the first series, the programme used an 'expert' panel of assessors (see Hardern 1954: 6–7). Thereafter, the format was reduced to the inventors, Hardern, Boumphrey and occasional expert assessors.
36 Mary Adams to Cecil McGivern, 9 March 1948, T32/216.
37 Ibid.
38 Ibid.
39 G. Boumphrey, L. Hardern, memorandum 'Inventors' Club', March 1953, T32/216/4.
40 24 February 1956; see Leonard Miall to D. Tel B, 'Productivity: Bouquet for Inventors' Club, 7 March 1956, T32/216.
41 G. Boumphrey, L. Hardern, memorandum 'Inventors' Club', March 1953, T32/216/4.
42 T32/303.
43 T32/248/1. Anon [Norman Collins] memorandum 'From the Controller of Television/Television Programmes', undated [late 1949/early 1950], 2, 4, R34/885/1.
44 *Passim*, T32/305/2.
45 Draft script preserved in T14/1117.
46 *Passim*, T14/802.
47 See T6/34.
48 Anon. [Norman Collins] memorandum 'From the Controller of Television/Television Programmes', undated [late 1949/early 1950]: 2, R34/885/1.
49 BBC staff lists.
50 McGivern to Rotha, 15 September 1952, LI 376.
51 McGivern to Director of Television Broadcasting, 'Matters Outstanding', 6 August 1952, T16/154/2.
52 His substitution of 'reality' for 'actuality'.
53 For the first of these he suggested five topics, three of which were scien-

tific or technological, on hydro-electricity, Teddington Research Laboratories and horology.

54 Duncan Ross to McGivern, 'Documentary Television and Programme Suggestions', 14 March 1948, T4/35.

55 Robert Barr to Controller, Television programmes, memo 'TV Documentary', 3 August 1951, T16/61/1.

56 Ibid. Elaine Bell argues that the context for this memorandum was that Barr was seeking to fight off an attempt by the Talks Department to take over the Documentary Unit (1986: 77). Equally, we may note that it was composed around the time when he was first applying for the Head of Department job in documentary, expressing his conclusion that documentary-drama was the appropriate form for television.

57 Undated memo in file marked 'documentaries – memos 1946–54', 'Selected list of documentary programmes written and produced by the Documentary Department, BBC Television Service', T4/35, supplemented by 'Documentary Department Output, April 1st 1954–March 1955', 1 March 1956, and 'Television Documentary Output', 21 May 1957, T16/61/2.

58 A valuable comparison can be made between this programme and the 1948 documentary film made by the Horizon Film Unit's *One Man's Story*, created under the same auspices as *The Centre* (see chapter 5).

59 J. Lyons (MOH, Todmorden) and C. W. Dixon (Department of Preventive Medicine and Public Health, University of Leeds), Lyons and Dixon 1953.

60 Barr to Rotha, 'Report on Programme, "Medical Officer of Health"', 25 March 1954, T4/34.

61 Audience Research Report, 5 October 1954, T4/34. For the measurement, see Silvey (1974: 116–17).

62 *Passim*, T32/322

63 See file T4/69.

64 McCall to CSA, 2 March 1953, LI 376.

65 The last appointment to the Department, A. G. Calder from Drama, occurred between October 1953 and October 1954 (BBC Staff List, January 1953).

66 Rotha, 'Television and the Future of Documentary', unpublished internal memo, 17 October 1954, quoted in Bell 1986: 71.

67 See 'Documentary Department Output April 1 1954–March 31 1955', T16/61/2.

68 See Anon. 1955. In fact, he was not a total stranger to the medium, as his film *Cover to Cover* had been the first to be shown on television in 1936.

69 *The World is Ours*, 30 June 1954, T32/346/6.

70 McGivern to Rotha, 22 December 1954, T16/61/1.
71 Paul Rotha, Annual Confidential Report, 22 April 1954, LI 376.
72 Emphasis added. Barnes to McGivern, 'Programmes', 19 March 1952, T16/154/ 2.

Chapter 7

1 Ian Jacob to Gerald Beadle (Director of Television Broadcasting), 20 July 1956, emphasis added, BBC Written Archives T16/61/2. (All archival references in this chapter are from this archive unless otherwise specified.)
2 McGivern to H. D. Tel and H. T. Tel., 'Documentary Programmes in Television', 5 March 1956, T16/61/2.
3 *Passim*, T16/61/2.
4 The limited survival of programmes from the earlier 1950s (for the technical reasons noted) prevents close analysis, but from the period under discussion it is possible to study some of the programmes themselves.
5 This greater informality of speech also began to influence filmmaking practice. For example, in the film *One Hundred Years Underground* (1963), the several people, including Herbert Morrison and John Betjeman, who speak on the history of the London 'tube', use an informal mode.
6 He had started at the BBC in 1948. Andrew Miller-Jones, previously the most active science producer, became preoccupied with *Panorama* from its launch in 1954.
7 For reference to *A Question of Science*, see Grace Wyndham Goldie to McGivern, 13 March 1958, T32/1820/1.
8 Goldie to McCloy, 17 January 1958, T32/1820/1; McCloy to Goldie, 'Science Programmes: Future Plans', 15 January 1958, ibid.
9 Goldie to Kenneth Adam, 'Science Output', 13 March 1958, T32/ 1820/1.
10 Ibid.
11 List in 'Science Broadcasts on Television', Appendix B to GAC 228, R6/239/1.
12 Adam had taken over from McGivern as Head of Television Production in February 1957 (see Briggs 1995: 15).
13 Goldie to Adam, 'DTel B's Quarterly Report for Board of Governors, April–Jul 1959', T16/310/2.
14 It was felt that 'Talks could and should handle the big controversial subjects in full length studies with film'; Memorandum, 'Television Documentary Output', 21 May 1957, T16/61/2.
15 I am grateful to Toni Charlton for this insight.
16 Singer to A Berkeley Smith (AHOBTel) 'Proposals for series of ten pro-

grammes on research starting week 46', 15 August 1957, T14/1502/1. The ambition to cover the arts was soon dropped.

17 James McCloy to Singer, 'O.B. Science: "Eye to the Future"' 16 August 1957, T14/1502/1.

18 Goldie to Singer, 'OB Series: Science', 22 August 1957, ibid.

19 McCloy to HT Tel, 'Science Programmes: Frontiers of Science', 18 September 1957, T32/626/1.

20 Singer to Kenneth Adam, 'Eye on Research: Planning, Preparations and Policy Considerations', 17 October 1958, T14/1503.

21 Gordon Rattray Taylor, 'Note on a meeting with Lord Adrian at Cambridge, October 29' [1957], T14/1502/1.

22 T14/1495-6.

23 L. E. Jeanes to Singer, 12 February 1958, T14/1502.

24 Head of OB, TV to DD Tel B, 'D Tel B's Quarterly Report for Board of Governors, January–March 1958', T16/310/1. They also managed to increase its 'comparative minority audience' by broadcasting in an earlier slot.

25 See file T16/1648.

26 Board of Management Minutes, 29 Sept 1959, which reports a meeting on 23 September 1958 between BBC representatives and leaders of the scientific community.

27 Worries about the coverage of the scientists' subject on television – which was newly a mass medium – certainly had causes other than merely professional *amour propre*. Given that this delegation excluded medical and health broadcasts from their concerns – which they made clear were in physics, chemistry and biology – (Harman Grisewood, 'Science Broadcasts', 1 October 1958, R51/967/1) we may speculate that this approach betokened a concern about differentials between the sciences, not only in television coverage, but also in their funding. All three of these individuals were chemists of one kind or another, but all the organisations they represented could be expected to have an interest in medical sciences.

28 Cyril Hinshelwood subsequently denied this, but it was the very clear implication that the BBC staff took from the meeting. See Annexe to General Advisory Council, 29 April 1959: 1–2, R51/967/1.

29 Board of Governors, Minutes 29 September 1958, R11/2.

30 Harman Grisewood, 'Science Broadcasts', 1 October 1958, R51/967/1.

31 Kenneth Adam to Singer, 'Television Scientific Programmes', 1 October 1958, T16/623.

32 Singer to Kenneth Adam, '*Eye on Research*: Planning, Preparations and Policy Considerations', 17 October 1958, T14/1502/1 (response to Mc-

Givern to Singer, 1 October 1958, T16/623).

33 McCloy to Adam, 'Presentation of Science by Television', 7 October 1958, T16/623.

34 Singer to Adam, '*Eye on Research*: Planning, Preparations and Policy Considerations', 17 October 1958, T14/1502/1.

35 C. D. Shaw, 'Science Broadcasts: notes of a meeting held on Wednesday, 22 October 1958, R51/967/1 and T16/623.

36 Ibid.; Singer to Adam, '*Eye on Research*', 17 October 1958, T14/1502/1.

37 C. D. Shaw, 'Science Broadcasts' (Memorandum), 22 October 1958, R51/961/1. The recruitment of producers for science programmes had been a subject of discussion in late 1958. Leonard Miall, Head of Talks, Television, 'emphasised the fact that for television the need was not so much for first-class scientists as for young men who were or would be first-class producers and had some scientific training'.

38 Board of Management Minutes, 3 November 1958, paragraph 507, excerpted in T16/623.

39 Solly Zuckerman and Hinshelwood had taken active parts in the discussion, 'Science Broadcasting', 2 April 1959 (G. A. C. 228) copy in M2/8/9; for discussion see Annexe to General Advisory Council, 29 April 1959: 1–2, R51/967/1.

40 The discussion also produced agreement to go public about the Royal Society's role in providing advice (which had been in place since 1942). This was prompted by the decision of Granada, the new Independent Television (ITV) broadcaster, formally to associate with the BA over sixth-form science broadcasts; R. D. Pendlebury, 'Science Broadcasts: notes of a meeting held on Tuesday, 5 May 1958', R51/967/1.

41 Harman Grisewood, 'The Needs of Science Broadcasting', 20 October 1960, R51/967/1.

42 The Shell Film Unit also made a tercentenary documentary for the Shell Film Unit, *A Light in Nature*.

43 19 July; BBC online catalogue (http://open.bbc.co.uk/catalogue/infax/, accessed 21 February 2007).

44 Martin to Daly, T14/1502/3.

45 These included the examples discussed in chapter five and his Shell Film Unit film released in this same year, 1959, on fighting the biological and social causes of infectious diseases, *Unseen Enemies*.

46 Other scientific organisations submitting evidence included the Association of Scientific Workers, the British Medical Association and the Medical section of the SFA, but these were not cited in the report's recommendations.

47 Excerpt from Report of the Committee on Broadcasting (Pilkington

Committee), R6/239/1.

48 Martin to Marriot, 15 October 1962, R6/239/1. Marriott described Fleck as 'one of the most persistent of our critics'; Marriott to Hood, 16 October 1962, R6/239/1.

49 Marriot to Head of Talks, 9 August, R6/239/1.

50 Marriot, 'Note of a Meeting at Burlington House, 12 December 1962', R6/239/1.

51 'Notes on Horizon Policy', unsigned, undated, c. February 1963, T14/3,316/1.

52 Daly to Singer, 'Horizon Magazine Programme', 5 March 1963, T14/3,316/1.

53 Singer to Daly, 'Horizon', 22 November 1963, T14/3,316/1.

54 Ramsay Short, 'Reasons for Calling the Meeting; Notes on Horizon Meeting Held 7 January 1964', T14/3,316/1.

55 See Michael Latham and Ramsay Short to Singer, T14/3,316/1.

56 For details of programmes see BBC online Programme Catalogue, http://open.bbc.co.uk/catalogue/infax/

57 Martin Freeth and David Dugan, private communication, February 2007.

58 Singer to D Tel, 'Science and Technology on BBC-1: A Re-appraisal', 23 November 1964, T16/623.

59 Glyn Jones to Peter Bruce, Peter Ryan and Raymond Baxter, '"Modern Age" – Programme Brief', 17 May 1965, T14/2935/1.

60 *Tomorrow's World* Audience Research Report, 5 August 1965, T14/2935/1.

61 McCloy advocated showmanship but stated that 'controversy is unfortunately rarely possible as a component to invite the attention of the large audience ... Where it is present it is usually at the frontier of knowledge and scientists arguing it out in a natural manner in front of cameras would be unintelligible to an audience with no background in science' (1963: 11).

62 For an analysis of this clause, see Silverstone (1985: 160–80).

Coda

1 http://open.bbc.co.uk/catalogue/infax/

2 See Frank Gillard, 'Nature Unit – West Region', November 1957, BBC Written Archives T16/736/2.

3 Young (1986) gives a full list of the programmes and his participant's account of what he saw as a dispiriting episode in which the radical and critical voice of science studies was squeezed out by the processes of

television and politics.

4 A joint organisation of the Royal Society, British Association and Royal Institution.

5 Adam Curtis, private communication, 13 April 2007.

BIBLIOGRAPHY

Archives consulted

Caversham, Reading: BBC Written Archives Centre.
Edinburgh: National Library of Scotland Manuscript Division: Peter Ritchie Calder Papers (Dep. 370).
Houston, Texas: Huxley, Julian Sorell Huxley – Papers, 1899–1980, MS 50, Woodson Research Center, Fondren Library, Rice University.
London: British Library of Political and Economic Science, Library Archives: Political and Economic Planning Archives.
London: National Archives, Kew: Records of Ministry of Information, Central Office of Information, Ministry of Health, Medical Research Council, Empire Marketing Board and General Post Office Film Unit, Ministry of Works, Ministry of Agriculture, Ministry of Food.
London: Royal Society: COPUS papers.
London, Victoria and Albert Museum, Royal Institute of British Architects Archive, Edward Carter Papers.
London: Wellcome Library: Eugenics Society Archive (SA/EUG).
Los Angeles, Department of Special Collections, Charles E. Young Research Library, University of California: Paul Rotha Papers (Collection 2001).
Norwich, University of East Anglia, Zuckerman Archive.
Reading: Isotype Collection, University of Reading Manuscript collection.
Stirling, University of Stirling: John Grierson Archive: letters from Paul Rotha to John Grierson.

Periodicals and reference sources

Amateur Photographer
Architects' Journal
BBC Staff Lists (Set at BBC Written Archives Centre)
British Medical Journal
Documentary News Letter
The Independent
The Lancet

The Listener
Oxford Dictionary of National Biography
Planning
Radio Times
Science and Film
The Times
World Film News
BBC Programme Catalogue http://catalogue.bbc.co.uk
British Film Institute Film & TV Database http://www.bfi.org.uk/filmtvinfo/ftvdb
British Film Institute Screenonline http://www.screenonline.org.uk

Works cited

Abel, R. (2005) *Encyclopedia of Early Cinema.* London: Routledge.
Adams, A. (1996) *Architecture in the Family Way: Doctors, Houses and Women, 1870–1900.* Montreal: McGill-Queen's University Press.
Aitken, I. (1990) *Film and Reform: John Grierson and the Documentary Film Movement.* London: Routledge.
____ (1998) *The Documentary Film Movement: An Anthology.* Edinburgh: Edinburgh University Press.
Altick, R. (1978) *The Shows of London.* Cambridge, Mass: Belknap Press.
Anker, P. (2001) *Imperial Ecology: Environmental Order in the British Empire.* Cambridge, MA: Harvard University Press.
____ (2005) 'The Bauhaus of Nature', *Modernism/Modernity*, 12, 229–51.
Anon. (1903a) 'Mr Martin Duncan on Photomicrography at the Camera Club', *The Amateur Photographer*, 176.
____ (1903b) 'Notes and Comments', *The Amateur Photographer*, 41.
____ (1903c) 'Notes', *Nature*, 27 August, 68, 394–7.
____ (1903d) 'Elevate the Standard of Your Picture Event. *The Unseen World* (advertisement)', *The Era*, 15 August, 27.
____ (1903e) 'Alhambra: New Grand Dramatic Ballet (listing)', *The Era*, 29 August, 18.
____ (1932) *The Film in National Life.* London: G. Allen & Unwin.
____ (1934a) 'The E. D. A. Film Campaign', *EDA Bulletin*, 3–6.
____ (1934b) 'The Committee Against Malnutrition', *The Lancet*, 226, 1358-1360.
____ (1935) 'Why We Want a National Food Policy', *Planning*, 44, 12 February, 1–2.
____ (1936a, supplement 1937) *Catalogue of British Medical Films.* London: British Film Institute.

____ (1936b) 'Mr. Therm – Film Star. Gas Light and Coke Company's Films', *Gas Journal*, 215, 325.
____ (1936c) 'A Review of the Gas Light and Coke Company's Films', *Gas Journal*, 216, 157–8.
____ (1936d) 'The Gas Light and Coke Company's Screen Publicity', *The Gas World*, 105, 195–6.
____ (1936e) 'The Documentary Film Consultative Body Formed', *The Times*, 10 January, 10.
____ (1936f) 'Plan for non-fiction films: Professors to act as advisors', *Kine Weekly*, 16 January, 22.
____ (1936g) 'Edgar Anstey joins March of Time', Daily Film Renter, 22 August, n.p.
____ (1936h) 'Whitehall defaults', *World Film News*, October, 29.
____ (1936i) 'Food on the Screen: Two Excellent Shorts', *Kine Weekly*, 8 October, 23.
____ (1936j) 'Films in Education. New Weekly Programme in London', *The Times*, 9 October, 8.
____ (1936k) 'A National Problem', *The Times*, 12 October, 8.
____ (1936l) '*Nutrition* at the Cinema', *The Lancet*, 17 October, 927.
____ (1936m) 'Malnutrition: Opposition Challenge (Medical notes in Parliament)', *British Medical Journal*, 24 October, 157–9.
____ (1936n) 'Medical News', *British Medical Journal*, 24 October, 849.
____ (1936o) '"*Nutrition*": Interesting Film in Grimsby', *Grimsby News*, 30 October (microfiche).
____ (1936p) 'Free "Talkie" Show: Gas Company's Interesting Film Venture', *Newcastle Journal*, 6 November (microfiche).
____ (1936q) 'Manchester Film Institute', *Manchester Guardian*, 30 November (microfiche).
____ (1936r) 'The Scientist Butts In', *Planning*, 88, 15 December, 1–3, 14.
____ (1936s) 'When Will Consumers Wake Up?', *Planning*, 89, 29 December, 1–2.
____ (1937a) *News for You about the New Book-Paper*. London: Fact.
____ (1937b) 'Untitled', *Today's Cinema*, 20 January, 14.
____ (1937c) 'Up and Down the Street', *Today's Cinema*, 1 February, 2.
____ (1937d) 'Monkeys and Children. Alec King Annoyed with *March of Time*', *Today's Cinema*, 12 February, 1.
____ (1937e) 'Whitehall Might Note', *World Film News*, March, 7.
____ (1937f) 'Documentary Films. A Panel of Lecturers', *The Times*, 30 August, 8.
____ (1938a) 'Food and National Health: Exhibition at Charing Cross Underground', *The Gas World*, 108, 97.

____ (1938b) 'Films for Scientists', *The Scientific Worker*, 46–7.
____ (1938c) 'Scientific Films', *The Scientific Worker*, 153–4.
____ (1938d) 'News', *Eugenics Review*, 30, 203–4.
____ (1939a) 'The Educational and Cultural Film', *The Scientific Worker*, 40–3.
____ (1939b) 'A Survey of Research', *Planning*, 156, 5 December, 1–2.
____ (1940a) 'Scientific Films', *The Scientific Worker*, xii, 46.
____ (1940b) 'Social Environment and the War', *Architects' Journal*, 1 February, 132.
____ (1940c) 'Housing as a Postwar Problem', *Architects' Journal*, 1 February, 156.
____ (1940d) 'PEP Work 1938–40', *Planning*, 166, 23 April, 2–9.
____ (1941a) 'Scientific Films', *The Scientific Worker*, 197–8.
____ (1941b) *The Scientific Film*. London: Scientific Films Committee.
____ (1941c) 'Rotha Films: A New Unit in Documentary', *Documentary News Letter*, 59.
____ (1941d) 'Research and its Application', *Planning*, 161, 13 February, 11–13.
____ (1943a) '35 Persons Killed in Air Crash', *The Times*, 22 January, 4.
____ (1943b) 'Major Eric Knight (Obituary)', *The Times*, 23 January, 6.
____ (1943c) 'Problems of the Screen', *The Times*, 3 June, 5.
____ (1944) 'A National Need', *Documentary News Letter*, 5, 3, August, 25.
____ (1946a) *The Classification, Appraisal and Grading of Scientific Films*. London: Scientific Film Association.
____ (1946b) *Catalogue of Films of General Scientific Interest available in Great Britain, etc.* London: Aslib, 188.
____ (1946c) '*Britain Can Make It* (Review)', *Monthly Film Bulletin*, 13, 145, 31 January, 11.
____ (1946d) '*Britain Can Make It* (Review)', *Documentary News Letter*, 6, 24.
____ (1947a) 'World Reconstruction: Development Dependent on Expanding Economy. Need for International Action', *The Times*, 23 August, 5.
____ (1947b) 'Edinburgh Festival: Documentary Films', *The Times*, 10 September, 7.
____ (1947c) 'United Nations Food Organisation: Sir John Boyd Orr First Director-General', *The Times*, 29 October, 3.
____ (1952) 'Across Frontiers: A Note on the International Scientific Film Association', *Science and Film*, 1, 3–5.
____ (1953a) *Supplement to the S.F.A. Catalogue of Medical Films*. London: Harvey & Blythe, v, 55.
____ (1953b) 'Questions for twins', *Radio Times*, 6 November, 15.
____ (1957) 'Star Quest', *Radio Times*, 19 April, 4.

____ (1958) 'Science is News', *Radio Times,* 10 October, 19.
____ (1992a) 'Arthur Calder-Marshall (Obituary)', *The Times,* 22 April, 13.
____ (1992b) 'Arthur Calder-Marshall (Obituary)', *The Independent,* 23 April, 29.
____ (2002) 'Gerald Harvey Thompson'. Available at: http://www.worldeducationalfilms.com/biogs/geraldthompson/, accessed 4 March 2007.
Anstey, E. (1982) 'Lord Ritchie Calder (letter)', *The Times,* 12 February, 12.
Anstey, E., G. Bell, M. Clarke and D. Ward (1963) 'Presentation Techniques'. *Journal of the Society of Film and Television Arts Limited,* Winter 1963–64, 7–10.
Armstrong, D. (1983) *Political Anatomy of the Body.* Cambridge: Cambridge University Press.
Arts Enquiry (1947) *The Factual Film.* Oxford: Oxford University Press.
Association of Scientific Workers (1952) *Graded List of Scientific Films.* London: Harvey and Blythe Ltd.
Baker, J. R. (1976) 'Julian Sorell Huxley', *Biographical Memoirs of Fellows of the Royal Society,* 22, 207–38.
Balfour, M. (1979) *Propaganda in War, 1939–1945: Organisations, Policies and Publics in Britain and Germany.* London: Routledge & Kegan Paul.
Barnes, B. (1977) *Interests and the Growth of Knowledge.* London: Routledge.
____ (1985) *About Science.* Oxford: Basil Blackwell.
Barnes, B. and D. Edge (1982) 'Part 5. Science as Expertise', in B. Barnes and D. Edge (eds), *Science in Context.* Milton Keynes: The Open University Press, 233–49.
Barron, A. L. E. (1965) 'Our Club: Some Highlights and Personalities of its First One Hundred Years', *Journal of the Quekett Microscopical Club,* 30, 61–8.
Bell, E. (1986) 'The Origins of British Television Documentary: The BBC 1946–55', in J. Corner (ed.) *Documentary and the Mass Media.* London: Edward Arnold, 65–80.
Bell, G. (1943) 'The Film and Science', *Scientific Worker,* 37–8.
____ (1963) 'Past, Present and Future', *Journal of the Society of Film and Television Arts Limited,* Winter, 2–7.
Berman, M. (1988) *All That is Solid Melts Into Air: The Experience of Modernity.* New York and London: Penguin.
Black, P. (1972) *The Mirror in the Corner : People's Television.* London: Hutchinson and Co.
Blaxter, K. L. (2004) 'Orr, John Boyd, Baron Boyd Orr (1880–1971), rev', *Oxford Dictionary of National Biography.* Oxford: Oxford University Press. Available at: http://www.oxforddnb.com/view/article/31519, accessed 5 December 2006.

Bleaney, B. (2006) 'Oliphant, Sir Marcus Laurence, Elwin (1901–2000)', *Oxford Dictionary of National Biography*. Oxford: Oxford University Press. Available at: http://www.oxforddnb.com/view/article/74397, accessed 12 September 2006.

Bloor, D. (1991) *Knowledge and Social Imagery*. Chicago: University of Chicago Press.

Bonney, T. G. and Rev. Y. Foote (2004) 'Duncan, Peter Martin (1824–1891)', *Oxford Dictionary of National Biography*. Oxford: Oxford University Press. Available at: http://www.oxforddnb.com/view/article/8228, accessed 24 May 2006.

Boon, T. (1990) '"Lighting the Understanding and Kindling the Heart?" Social Hygiene and Propaganda Film in the 1930s. (Précis of paper given to annual conference of the Society for the Social History of Medicine, 1989)', *Social History of Medicine*, 3, 140–1.

____ (1993) 'The Smoke Menace: Cinema, Sponsorship, and the Social Relations of Science in 1937', in M. Shortland (ed.) *Science and Nature (BSHS Monograph 8)*. Oxford: BSHS, 57–88.

____ (1997) 'Agreement and Disagreement in the Making of "World of Plenty"', in D. Smith (ed.) *Nutrition in Britain: Science, Scientists and Politics in the Twentieth Century*. London: Routledge, 166–89.

____ (1999) 'Films and the Contestation of Public Health in Interwar Britain', unpublished doctoral thesis, University of London.

____ (2000) '"The Shell of a Prosperous Age": History, Landscape and the Modern in Paul Rotha's *The Face of Britain* (1935)', in C. Lawrence and A. Mayer (eds) *Regenerating England: Science, Medicine and Culture in the Interwar Years*. Amsterdam: Rodopi, 107–48.

____ (2004a) 'Industrialisation and Catastrophe: The Victorian Economy in British Film Documentary, 1930–1950', in M. Taylor and M. Wolff (eds) *The Victorians Since 1901: Histories, Representations and Revisions*. Manchester: Manchester University Press, 107–20.

____ (2004b) 'Science and the Citizen: Documentary Film and Science Communication in World War Two', in R. Fox and I. Maclean (eds) *History and the Public Understanding of Science*. Available at: http://www.mfo.ac.uk/Publications/actes2/boon.htm, accessed 14 December 2006.

____ (2005) 'Health Education Films in Britain, 1919–1939: Production, Genres and Audiences', in G. Harper and A. Moor (eds) *Signs of Life: Cinema and Medicine*. London: Wallflower Press, 45–57.

____ (2008) 'Disease Narratives and Tuberculosis Health Education in Britain, 1930–1960', in F. Condrau and M. Worboys (eds) *From Urban Penalty to Global Emergency*. Montreal: McGill-Queen's University Press.

Bottomore, S. (2005) 'Pike, Oliver', in R. Abel (ed.) *Encyclopedia of Early Cin-*

ema. London: Routledge, 521.

Bowler, P. (2006) 'Presidential Address: Experts and Publishers: Writing Popular Science in Early Twentieth-Century Britain, Writing Popular History of Science Now'. *British Journal for the History of Science* 39, 159–87.

Bragg, L. (1960) 'Popularizing Science on Television', *The Listener*, 14 January, 75–6.

Branson, N. and M. Heinemann (1973) *Britain in the Nineteen Thirties*. London: Weidenfeld and Nicolson.

Brey, P. (2003) 'Theorizing Modernity and Technology', in T. J. Misa, P. Brey and A. Feenberg. Cambridge, Mass: MIT Press, 33–72.

Briggs, A. (1979) *Sound and Vision: The History of Broadcasting in the United Kingdom vol IV*. Oxford: Oxford University Press.

____ (1995) *Competition: The History of Broadcasting in the United Kingdom vol V*. Oxford: Oxford University Press.

British Commercial Gas Association (1939) *Modern Films on Matters of Moment*. London: British Commercial Gas Association.

Broks, P. (1996) *Media Science Before the Great War*. Basingstoke: Macmillan.

Brome, V. (1988) *J. B. Priestley*. London: Hamish Hamilton.

Brown, A. (2005) *J. D. Bernal, The Sage of Science*. Oxford: Oxford University Press.

Brown, G. and L. Enticknap (2003) 'Jackson, Pat (1916–)', *Screenonline*. Available at: http://www.screenonline.org.uk/people/id/513088/index.html, accessed 8 January 2007.

Burns, J. (2000) 'Watching Africans Watch Films: Theories of Spectatorship in British Colonial Africa', *Historical Journal of Film, Radio and Television*, 20, 197–211.

Burt, J. (2002) *Animals in Film*. London: Reaktion.

Cahan, D. (ed.) (2003a) *From Natural Philosophy to the Sciences: Writing the History of Nineteenth-Century Science*. Chicago and London: University of Chicago Press.

____ (2003b) 'Institutions and Communities', in D. Cahan (ed.), *From Natural Philosophy to the Sciences: Writing the History of Nineteenth-Century Science*. Chicago and London: University of Chicago Press, 291–328.

Calvocoressi, P., G. Wint, J. Pritchard (1989) *Total War: The Causes and Courses of the Second World War*. London: Penguin.

Cannadine, D. (1984) 'The Present and the Past in the English Industrial Revolution'. *Past and Present*, 103, 131–72.

Cantor, G., G. Dawson, G. Gooday, R. Noakes, S. Shuttleworth and J. Topham (eds) (2004) *Science in the Nineteenth-Century Periodical*. Cambridge: Cambridge University Press.

Carlson, W. B. (1997) 'Innovation and the Modern Corporation: From Heroic

Invention to Industrial Science', in J. Krige and D. Pestre (eds) *Science in the Twentieth Century*. Amsterdam: Harwood Academic Publishers, 203–26.
Carpenter, H. (1997) *The Envy of the World: Fifty Years of the Third Programme and Radio Three*. London: Weidenfeld & Nicolson.
Carroll, V. (2004) 'The Natural History of Visiting: Responses to Charles Waterton and Walton Hall', *Studies in History and Philosophy of Biological and Biomedical Sciences*, 35C, 31–64.
Cartwright, N., J. Cat, L. Fleck and T. Uebel (1996) *Otto Neurath: Philosophy Between Science and Politics*. Cambridge: Cambridge University Press.
Cawston, R. (1958) 'On Call to a Nation', *Radio Times*, 17 October, 9.
Central Information Bureau for Educational Films Ltd (ed.) *A National Encyclopaedia of Educational Films and 16mm Apparatus Available in Great Britain*. London: C.I.B.E.F.
Chanan, M. (1996) *The Dream that Kicks: The Prehistory and Early Years of Cinema in Britain*, second edn. London: Routledge.
Chaney, D. and M. Pickering (1986) 'Authorship in Documentary: Sociology as an Art Form in Mass Observation', in J. Corner (ed.) *Documentary and the Mass Media*. London: Edward Arnold, 29–44.
Chapman, J. (1998) *The British at War: Cinema, State and Propaganda, 1939–45*. London: I.B. Tauris.
Charney, L. and V. R. Schwartz (eds) (1995) *Cinema and the Invention of Modern Life*. Berkeley: University of California Press.
____ (1995) 'Introduction', in L. Charney and V. R. Schwartz (eds) (1995) *Cinema and the Invention of Modern Life*. Berkeley: University of California Press, 1–12.
Chibnall, B. (ed.) (1966) *The British Film Guide, Vol 1: Medicine and Allied Subjects*. London: British Film Institute.
Chibnall, B. and J. Le Harivel (1960) 'Report on Meetings of the Popular Science Film Section', *Science and Film*, 8, 40–4.
Childs, P. (2000) *Modernism*. London: Routledge.
Christie, I. (1991) 'Making Sense of Early Soviet Sound', in R. Taylor and I. Christie (eds) *Inside the Film Factory: New Approaches to Russian and Soviet Cinema*. London: Routledge, 176–92.
____ (1994) *The Last Machine: Early Cinema and the Birth of the Modern World*. London: BBC Educational Developments.
____ (1998) 'The Avant-Gardes and European Cinema Before 1930', in J. Hill and P. C. Gibson (eds) *The Oxford Guide to Film Studies*. Oxford: Oxford University Press, 449–54.
Clark, R. (1968) *J.B.S., the Life and Work of J.B.S. Haldane*. Oxford: Oxford University Press.
Clarke, M. (1959) 'Science International: A Review of Two Recent BBC

Television Programmes', *Scientific Film Review*, 5, 303–6.
Clarke, P. (2002) *The Cripps Version: The Life of Sir Stafford Cripps*. Allen Lane.
Cockburn, S. (1981) *Oliphant: The Life and Times of Sir Mark Oliphant*. Adelaide: Axiom.
Collins, H. (1987) 'Certainty and the Public Understanding of Science: Science on Television', *Social Studies of Science*, 17, 689–713.
Collins, P. (1981) 'The British Association as Public Apologist for Science, 1919–1946', in R. Macleod and P. Collins (eds) *The Parliament of Science: The BAAS, 1831–1981*. London: Science Reviews Ltd, 211–36.
Colls, R. (1986) 'Englishness and the Political Culture', in R. Colls and P. Dodd (eds) *Englishness: Politics and Culture 1880–1920*. London: Croom Helm, 29–61.
Colls, R. and Dodd, P. (1985) 'Representing the Nation: The British Documentary Film, 1930–45', *Screen*, 26, 21–33.
Cook, D. A. (1981) *A History of Narrative Film*. New York: W. W. Norton & Company.
Cook, P. (ed.) (1985) *The Cinema Book*. London: British Film Institute.
Cooper, F. (2005) *Colonialism in Question: Theory, Knowledge, History*. Berkeley, CA: University of California Press.
Coopey, R., N. Tiratsoo and S. Fielding (1993) *The Wilson Governments 1964–1970*. London and New York: Pinter Publishers.
Cooter, R. and S. Pumfrey (1994) 'Separate Spheres and Public Places: Reflections on the History of Science Popularisation and Science in Popular Culture', *History of Science*, 32, 237–67.
Corke, H. (1960a) 'Critic on the Hearth', *The Listener*, 7 April, 636.
____ (1960b) 'Critic on the Hearth', *The Listener*, 14 April, 680.
Corner, J. (ed.) (1986) *Documentary and the Mass Media*. London: Edward Arnold.
____ (1991a) 'General Introduction: Television and British Society in the 1950s', in J. Corner (ed.), *Popular Television in Britain: Studies in Cultural History*. London: British Film Institute, 1–21.
____ (1991b) 'Documentary Voices', in J. Corner (ed.), *Popular Television in Britain: Studies in Cultural History*. London: British Film Institute, 42–59.
Coxhead, E. (1936) 'The Nutrition Film'. *Left Review*, 2, 777.
Cripps, R. S. (1948) 'The Documentary Film', in P. Noble (ed.), *British Film Yearbook 1947–48*. London: Skelton Robinson, 64–6.
Crothall, G. (1999) 'Images of Regeneration: Film Propaganda and the British Slum Clearance Campaign, 1933–1938', *Historical Journal of Film, Radio and Television*, 19, 339–58.
Crowther, J. (1970) *Fifty Years with Science*. London: Barrie & Jenkins.

Crowther, J. G., O. J. R. Howarth and D. P. Riley (eds) (1942) *Science and World Order*. Harmondsworth: Penguin.

Curran, J. and V. Porter (eds) (1983) *British Cinema History*. London: Wiedenfeld and Nicolson.

Dale, H. (1950) 'Science and Broadcasting', *BBC Quarterly*, 5, 136–41.

Daly, P. (1964) 'Horizon', *Radio Times*, 30 April, 8.

Darling, E. (2000) '"Enriching and Enlarging the Whole Sphere of Human Activities": The Work of the Voluntary Sector in Housing Reform in Inter-War Britain', in C. Lawrence and A. Mayer (eds) *Regenerating England: Science, Medicine and Culture in the Interwar Years*. Amsterdam: Rodopi, 149–78.

Davies, G. (2000a) 'Narrating the Natural History Unit: Institutional Orderings and Spatial Strategies', *Geoforum*, 31, 539–51.

____ (2000b) 'Science, Observation and Entertainment: Competing Visions of Postwar British Natural History Television', *Ecumene*, 7, 432–60.

De Mouilpied, H. (1937a) 'Films in the Making', *Co-Partners' Magazine (Gas Light and Coke Co.)*, 27, 408.

____ (1937b) 'The Gas Industry Presents', *Co-Partners' Magazine (Gas Light and Coke Co.)*, 27, 610–11.

Desmarais, R. (2004) '"Promoting Science": The BBC, Scientists, and the British Public, 1930–1945', unpublished MSc thesis, University of London.

Dimmock, P. (1951) 'Television Out and About'. *BBC Year Book 1951*. London: BBC.

Divall, C. (1992) 'From a Victorian to a Modern: Julian Huxley and the English Intellectual climate', in C. K. Waters and A. Van Helden (eds) *Julian Huxley: Biologist and Statesman of Science*. Houston: Rice University Press, 31–44.

Dodd, K. and P. Dodd (1996) 'Engendering the Nation: British Documentary Film, 1930–1939', in A. Higson (ed.) *Dissolving Views: Key Writings on the British Cinema*. London: Cassell, 38–50.

Dodd, P. (1986) 'Englishness and the National Culture', in R. Colls and P. Dodd (eds) *Englishness: Politics and Culture 1880–1920*. London: Croom Helm, 1–28.

Donald, J., A. Friedberg and L. Marcus (eds) (1999) *Close Up 1927–1933: Cinema and Modernism*. Princeton: Princeton University Press.

Donaldson, L. (1912) *The Cinematograph and Natural Science: The Achievements and Possibilities of Cinematography as an Aid to Scientific Research*. London: Ganes Ltd.

Doncaster, C. (1956) 'The Story Documentary', in P. Rotha, *Television in the Making*. London: Focal Press, 44–8.

Drouin, J.-M. and B. Bensaude-Vincent (1996) 'Nature for the People', in N. Jardine, J. Secord and E. Spary (eds) *Cultures of Natural History*. Cambridge: Cambridge University Press, 408–25.

Duncan, F. M. (1902) *First Steps in Photo-Micrography: A Handbook for Novices*. London: Hazell, Watson & Viney.

____ (1903) 'Worms, and How to Photograph Them', *The Amateur Photographer*, 38, 48–9.

____ (1921) 'Some Methods of Preparing Marine Specimens', *Journal of the Quekett Microscopical Club*, 2nd Series, XIV, 215–20.

Durbin, E. (1985) *New Jerusalems: The Labour Party and the Economics of Democratic Socialism*. London: Routledge & Kegan Paul.

Durden, J., M. Field and P. Smith (1941) *Cine-Biology*. Harmondsworth: Penguin Books.

Eden, W. (1935) 'The English Tradition in the Countryside', *Architectural Review*, LXXVII, 85–94, 193–202, 142–52.

Edwards, J. and M. Twyman (1975) *Graphic Communication Through ISOTYPE*. Reading: University of Reading.

Eisenstein, S., V. Pudovkin and G. Alexandrov (1994 [1928]) 'Statement on Sound', in R. Taylor and I. Christie (eds) *The Film Factory: Russian and Soviet Cinema in Documents 1896–1939 (1928)*, second edn. London: Routledge, 234–5.

Ellis, J. (2000) *John Grierson: Life, Contributions, Influence*. Carbondale, IL: Southern Illinois University Press.

Elzinga, A. and C. Landström (eds) (1996) *Internationalism and Science*. Cambridge: Taylor Graham.

Enticknap, L. D. G. (1999) 'The Non-Fiction Film in Postwar Britain', unpublished PhD thesis, University of Exeter.

Essex-Lopresti, M. (1997) 'Centenary of the Medical Film', *The Lancet*, 349, 819–20.

____ (1998a) 'The Medical Film 1897–1997: Part One: The First Half Century', *Journal of Audiovisual Media in Medicine*, 21, 7–12.

____ (1998b) 'The Medical Film 1897–1997: Part Two. The Second Half-Century', *Journal of Audiovisual Media in Medicine*, 21, 48–55.

____ (1998c) 'Medicine and Televison', *Journal of Audiovisual Media in Medicine*, 20, 61–4.

Field, A. M., J. V. Durden and F. P. Smith (1952) *See How They Grow. Botany Through the Cinema*. Harmondsworth: Penguin Books.

Field, M. (1943) 'The Making of Biological Films', *Advancement of Science*, 2, 303–5.

Field, M. and P. Smith (1934) *Secrets of Nature*. London: Faber & Faber.

Fielding, R. (1978) *The March of Time, 1935–1951*. New York: Oxford University Press.

Fish, R. (1976) 'The Library and Scientific Publications of the Zoological Society of London', in S. Zuckerman (ed.) *The Zoological Society of London, 1826–1976 and Beyond.* London: Academic Press for the Zoological Society of London, 233–52.

Forbes, A. (1954) 'Spotlight on Inventors' Club', *Unknown magazine, copy in BBCWAC T32/216/4.*

Fox, J. (2005) 'John Grierson, his "Documentary Boys" and the British Ministry of Information, 1939–1942', *Historical Journal of Film, Radio and Television,* 25, 345–69.

Francis, M. (1999) 'The Labour Party: Modernisation and the Politics of Restraint', in B. Conekin, F. Mort and C. Waters (eds) *Moments of Modernity: Reconstructing Britain: 1945-1964.* London: Rivers Oram, 152–70.

Frayling, C. (1995) *Things to Come.* London: British Film Institute.

____ (2005) *Mad, Bad and Dangerous: The Scientist and the Cinema.* London: Reaktion.

Galison, P. (1990) 'Aufbau/Bauhaus: Logical Positivism and Architectural Modernism', *Critical Enquiry,* 16, 709–52.

Gardner, C. and R. M. Young (1981) 'Science on TV: A Critique', in T. Bennett, S. Boyd-Bowmann, C, Mercer and J. Woollacott. *Popular Television and Film.* London: British Film Institute, 171–93.

Gaycken, O. (2002) "A Drama Unites Them in a Fight to the Death': Some Remarks on the Flourishing of a Cinema of Scientific Vernacularisation in France, 1904–1914', *Historical Journal of Film, Radio and Television,* 22, 353–74.

____ (2003) 'The Sources of The Secrets of Nature: The Popular Science Film at Urban, 1903–1911', in A. Burton and L. Porter (eds) *Scene-Stealing: Sources for British Cinema Before 1930.* London: Flicks Books, 36–42.

____ (2005) 'Devices of Curiosity: Cinema in the Field of Scientific Visuality', unpublished PhD thesis, University of Chicago.

Gibson, C. R. (1906) *The Romance of Modern Electricity.* London: Seeley & Co.

Giddens, A. (1991) *Modernity and Self-Identity: Self and Society in the Late Modern Age.* Palo Alto: Stanford University Press.

Gold, J. R. (1997) *The Experience of Modernism : Modern Architects and the Future City, 1928–53.* London: Routledge.

Goodchild, P. (2004) 'Clouds on the Horizon', *The Guardian,* 7 October, http://www.guardian.co.uk/life/feature/story/0,,1320986,00.html, accessed 22 February 2007.

Gordon, D. (1999) *A History of the Shell Film & Video Unit 1934 to 1999.* London: Shell.

Gransden, K. (1958) 'Critic on the Hearth', *The Listener*, 30 October, 704.
Grant, B. K. (2006) *Film Genre: From Iconography to Ideology*. London: Wallflower Press.
Grant, M. (1990) 'The National Health Campaigns of 1937–1938', in D. Fraser (ed.) *Cities, Class and Communication: Essays in Honour of Asa Briggs*. New York: Harvester Wheatsheaf, 216–33.
____ (1994) *Propaganda and the Role of the State in Inter-War Britain*. Oxford: Oxford University Press.
____ (1999) 'Towards a Central Office of Information: Continuity and change in British Government Information policy, 1939–51', *Journal of Contemporary History*, 34, 49–67.
Grattan, D. (2002) 'James McCloy (Obituary)', The Independent, 3 August, (microfiche).
Greene, G. (1995) *Mornings in the Dark: The Graham Greene Film Reader*, ed. D. Parkinson. Manchester: Carcanet.
Gregory, J. and S. Miller (1998) *Science in Public: Communication, Culture and Credibility*. New York: Plenum.
Gregory, R. (ed.) (1946) 'The Dissemination of Scientific Information to the Public: Report of a conference held July 8, 1946'. *Advancement of Science*, 4, 19–36.
Gregory, S. (1941) 'Mr Arliss and Mr Wells', *The Scientific Worker*, 13, 186–7.
Grierson, J. (1954) 'The BBC and All That', *Quarterly of Film, Radio and Television*, IX, 46–59.
____ (1966) *Grierson on Documentary*, ed. F. Hardy. London: Faber and Faber.
Grindon, L. H. (1859) *Manchester Walks and Wild Flowers*. London.
Gunning, T. (2004) 'An Aesthetic of Astonishment: Early Film and the (In) Credulous Spectator', in L. Braudy and M. Cohen (eds) *Film Theory and Criticism: Introductory Readings*, sixth edn. New York: Oxford University Press, 862–76.
____ (2005) 'Modernity and Early Cinema', in R. Abel (ed.) *Encyclopedia of Early Cinema*. London: Routledge, 439–42.
Haggith, T. (1990) 'Postwar Reconstruction as Depicted in Official British Films of the Second World War', *Imperial War Museum Review*, 34–45.
____ (1998) '"Castles in the Air": British Film and the Reconstruction of the Built Environment, 1939-51', unpublished PhD thesis, University of Warwick.
Haldane, J. B. S. (1927) *Possible Worlds, and Other Essays*. London: Chatto & Windus.
____ (1938) *The Marxist Philosophy and the Sciences*. London: G. Allen & Unwin.

____ (1941) *Science and Everyday Life.* Harmondsworth: Penguin.
Hall, S. (1972) 'The Social Eye of *Picture Post*', *Working Papers in Cultural Studies, Birmingham Centre for Cultural Studies*, 2, 71–120.
Hammerton, J. (1999) 'Cheese-Mites, Mosses and Mould: Percy Smith and the Secrets of Nature', *Viewfinder*, 9–11.
Hannah, L. (1979) *Electricity before Nationalisation.* London: Macmillan.
Haraway, D. (1984) 'Teddy Bear Patriarchy: Taxidermy in the Garden of Eden, New York City, 1908–1936', *Social Text*, 11, 20–64.
Hardern, L. (1954) *TV Inventors' Club.* London: Rockcliff.
Harding, A. (2004) 'The Closure of the Crown Film Unit in 1952: Artistic Decline or Political Machinations?', *Contemporary British History*, 18, 22–51.
Harding, C. and S. Popple (eds) (1996) *In the Kingdom of Shadows: A Companion to Early Cinema.* London: Cygnus Arts.
Hardy, F. (ed.) (1979) *John Grierson: A Documentary Biography.* London: Faber and Faber.
Harris, J. (1992) 'War and Social History: Britain and the Home Front during the Second World War', *Contemporary European History*, 1, 17–35.
Harrison, F. (1934) 'Eyes and No Eyes', *Architectural Review*, 75, 188–92.
Hasegawa, J. (1999) 'The Rise and Fall of Radical Reconstruction in 1940s Britain', *Twentieth Century British History*, 10, 137–61.
Hayward, S. (1996) *Key Concepts in Cinema Studies.* London: Routledge.
Henderson, H. S. (1982) 'F. Percy Smith and the Development of Cine-Photomicrography', *Microscopy*, 34, 422–3.
Hewer, H. R. (1946) 'The Production and Use of Films for Teaching', in *Scientific Films Association: The Place of The Film in Medical Education.* London: Scientific Film Association, 16–17.
Higson, A. (1995) *Waving the Flag: Constructing a National Cinema in Britain.* Oxford: Clarendon.
Hilgartner, S. (1990) 'The Dominant view of Popularisation: Conceptual Problems, Political Uses', *Social Studies of Science*, 20, 519–39.
Hill, J. and P. Church Gibson (1998) *The Oxford Guide to Film Studies.* Oxford: Oxford University Press.
Hilton, M. (2006) 'Hilton, John (1880–1943)', *Oxford Dictionary of National Biography.* Oxford: Oxford University Press. Available at: http://www.oxforddnb.com/view/article/40915, accessed 20 December 2006.
Hodgkin, D. (1980) 'John Desmond Bernal'. *Biographical Memoirs*, 26, 17–84.
Hogben, L. (1936a) 'The New Visual Culture', *Sight and Sound*, 5, 6–9.
____ (1936b) *The Retreat from Reason.* London: Watts & Co.
____ (1937) 'Naturalistic Studies in the Education of the Citizen', in J. B. Orr

What Science Stands For. London: G. Allen & Unwin, 111–32.
____ (1938) *Science for the Citizen*. London: G. Allen & Unwin Ltd.
____ (1998) *Lancelot Hogben, Scientific Humanist: An Unauthorized Autobiography*. Woodbridge: Merlin.
Hogenkamp, B. (1986) *Deadly Parallels: Film and the Left in Britain, 1929–39*. London: Lawrence and Wishart.
Horner, D. (1993) 'The Road to Scarborough: Wilson, Labour and the Scientific Revolution', in R. Coopey, N. Tiratsoo and S. Fielding (eds) *The Wilson Governments 1964–1970*. London and New York: Pinter Publishers, 48–71.
Hughes, D. E. P. (2004) 'Martin, Sir David Christie (1914–1976)', *Oxford Dictionary of National Biography*. Oxford: Oxford University Press. Available at: http://www.oxforddnb.com/view/article/31417, accessed 14 September 2006.
Hughes, J., P. J. Martin and W. W. Sharrock (eds) (1995) *Understanding Classical Sociology (Marx, Weber, Durkheim)*. London: Sage.
Huxley, J. (1931) *Africa View*. London: Chatto and Windus.
____ (1934a) *Scientific Research and Social Needs*. London: Watts & Co.
____ (1934b) 'Secrets of Nature', *Sight and Sound*, 3, 120–1.
____ (1935) 'Making and Using Nature Films', *The Listener*, 13, 595.
____ (1936a) 'Eugenics and Society', *Eugenics Review*, 28, 11–31.
____ (1936b) 'Science and the Problems of Society', *The Human Factor*, 10, 125–8.
____ (1936c) 'All Party Planning: Its Necessity and Possibility', *Service in Life and Work*, 16–23.
____ (1941) *The Uniqueness of Man*. London: Chatto and Windus.
____ (1944) *On Living in a Revolution*. London: Chatto & Windus.
Jackson, K. (2004) *Humphrey Jennings*. London: Picador.
Joad, C. E. M. (1940) 'An Open Letter to H.G. Wells', *New Statesman and Nation*, 154–5.
Johnson, I. (2005) 'A City Speaks: The Story of Manchester's Municipal Musical', *Manchester Sounds*, 6, 109–31.
Jones, A. M. (1949) 'President's Address: History and Reminiscences of the Quekett Microscopical Club', *Journal of the Quekett Microscopical Club*, 4th Series, Vol 3, 1–10.
Jones, G. (1986) *Social Hygiene in 20th Century Britain*. London: Croom Helm.
Jones, S. G. (1987) *The British Labour Movement and Film, 1918–1939*. London: Routledge & Kegan Paul.
Jordanova, L. (1990) 'Medicine and Visual Culture (Review Article)', *Social History of Medicine*, 3, 89–99.

Karpf, A. (1988) *Doctoring the Media: The Reporting of Health and Medicine.* London: Routledge.
Kay, H. D. (1972) 'John Boyd Orr, Baron Boyd Orr of Brechin Mearns, 1880–1971', *Biographical Memoirs of Fellows of the Royal Society,* 18, 43–81.
Kevles, D. (1992) 'Huxley and the Popularization of Science', in C. K. Waters and A. Van Helden (eds) *Julian Huxley: Biologist and Statesman of Science.* Houston: Rice University Press, 238–51.
Knight, E. (1940) *Now Pray We for Our Country.* London: Cassell and Co.
____ (1941) *This Above All.* London: Cassell and Co.
____ (1952) *Portrait of a Flying Yorkshireman: Letters from Eric Knight in the United States to Paul Rotha in England.* London: Chapman and Hall.
Knight, E. and P. Rotha (1945) *World of Plenty: the Book of the Film.* London: Nicholson & Watson.
Kracauer, S. (1997) *Theory of Film: The Redemption of Physical Reality.* Princeton, NJ: Princeton University Press.
Kuhn, T. (1962) *The Structure of Scientific Revolutions.* Chicago: University of Chicago Press.
Landecker, H. (2006) 'Microcinematography and the History of Science and Film', *Isis,* 97, 121–32.
Lankford, J. (1981) 'Amateurs Versus Professionals: The Controversy over Telescope Size in Late Victorian Science', *Isis,* 72, 11–28.
Latour, B. (1987) *Science in Action: How to Follow Scientists and Engineers Through Society.* Milton Keynes: Open University Press.
Lawrence, C. (1998) 'Still Incommunicable: Clinical Holists and Medical Knowledge in Interwar Britain', in C. Lawrence and G. Weisz (eds) *Greater than the Parts: Holism in Biomedicine.* New York: Oxford University Press, 94–111.
Lawrence, G. (1990) 'Object Lessons in the Museum Medium', *New Research in Museum Studies, 1: Objects of Knowledge,* 103–24.
Lebas, E. (1995) '"When Every Street Became a Cinema": the Film Work of the Bermondsey Borough Council's Public Health Department', *History Workshop Journal,* 42–66.
Le Corbusier (1927) *Towards a New Architecture.* London: Architectural Press.
Le Harivel, J. (1960) 'Looking Ahead: TV and Film', *Scientific Film Scientifique,* 1, 11–14.
Le Mahieu, D. (1992) 'The Ambiguity of Popularization', in C. K. Waters and A. Van Helden (eds) *Julian Huxley: Biologist and Statesman of Science.* Houston: Rice University Press, 252–6.
____ (1988) *A Culture for Democracy: Mass Communication and the Cultivated Mind in Britain Between the Wars.* Oxford: Clarendon Press.

Leslie, S. C. (1937) 'Kensal House: The Case for Gas is Proved', *The Gas World*, 108, 299–304.
____ (1938) 'Letter to the Editor', *The Times*, 16 June, 12.
Leventhal, F. (ed.) (1995) *Twentieth Century Britain: An Encyclopedia.* New York: Garland Publishing, Inc.
Levy, D. H. (1997) 'England's Ultimate Amateur', *Sky and Telescope*, 93, 106–7.
Lewis, J. and B. Brookes (1983) 'A Reassesment of the work of the Peckham Health Centre', *Millbank Memorial Fund Quarterly*, 61, 307–50.
Leyda, J. (1964) *Films Beget Films.* London: George Allen & Unwin.
Lindsay, J. (1947) 'Documentaries of the Year', in *Informational Film Year Book.* Edinburgh: The Albyn Press, 33–42.
Lindsay, K. (1981) 'PEP Through the 1930s: Organisation, Structure, People', in J. Pinder (ed.) *Fifty Years of Political and Economic Planning: Looking Forward 1931–1981.* London: Heinemann, 9–31.
Lloyd, B. (ed.) (1948) *Science in Films: A World Review and Reverence Work.* London: Sampson Low, Marston & Co.
Loughlin, K. (2000) '"Your Life in Their Hands": The Context of a Medical-Media Controversy', *Media History*, 6, 177–88.
Low, A. M. (1954) *Thanks to Inventors.* London: Lutterworth.
Low, R. (1971) *The History of the British Film 1918–1929.* London: Routledge.
____ (1979a) *Documentary and Educational Films of the 1930s.* London: George Allen and Unwin.
____ (1979b) *Films of Comment and Persuasion of the 1930s.* London: George Allen and Unwin.
Luckin, B. (1990) *Questions of Power: Electricity and Environment in Inter-War Britain.* Manchester: Manchester University Press.
Lyons, J. and C. W. Dixon (1953) 'Smallpox in the Industrial Pennines, 1953', *Medical Officer*, 90, 293– 300.
Mackenzie, D. and J. Wajcman (1999) 'Introductory Essay: The Social Shaping of Technology', in D. Mackenzie and J. Wajcman (eds) *The Social Shaping of Technology*, second edn. Buckingham: Open University Press, 3–27.
MacNicol, J. (1980) *The Movement for Family Allowances: A Study in Social Policy Development.* London: Heinemann.
Macpherson, D. (ed.) (1980) *British Cinema: Traditions of Independence.* London: British Film Institute.
Maddison, J. (1948) 'The Scientific Film', *Penguin Film Review*, 7, 117–24.
Marris, P. (ed.) (1982) *Paul Rotha.* London: British Film Institute.
Marwick, A. (1964) 'Middle Opinion in the Thirties: Planning, Progress and Political Agreement', *English Historical Review*, 79, 285–98.

Matless, D. (1998) *Landscape and Englishness.* London: Reaktion.
Mayer, A.-K. (2003) 'Setting up a Discipline, II: British History of Science and the "End of Ideology", 1931–1948', *Studies in the History and Philosophy of Science,* 35, 41–72.
Mayhew, M. (1988) 'The 1930s Nutrition Controversy', *Journal of Contemporary History,* 23, 445–64.
McCloy, J. (1956) 'Frontiers of Science', *Radio Times,* 13 January, 27.
____ (1958a) 'Five Hundred Million Years', *Radio Times,* 18 April, 3.
____ (1958b) 'Science is News', *Radio Times,* 10 October, 9.
____ (1963) 'Science on Television', *Journal of the Society of Film and Television Arts Limited,* Winter, 11–13.
McGivern, C. (1950) 'The Big Problem', *BBC Quarterly,* 5, 142–9.
McGucken, W. (1979) 'The Social Relations of Science: The British Association for the Advancement of Science, 1931–1946', *Proceedings of the American Philosophical Society,* 123, 237–64.
____ (1984) *Scientists, Society, and State: The Social Relations of Science Movement in Great Britain, 1931–1947.* Columbus: Ohio State University Press.
McKernan, L. (2003) '"Something More than a Mere Picture Show": Charles Urban and the Early Non-Fiction Film in Great Britain and America, 1897–1925', unpublished PhD thesis, University of London.
____ (2004a) 'Smith, (Frank) Percy (1880–1945)', *Oxford Dictionary of National Biography.* Oxford: Oxford University Press. Available at: http://www.oxforddnb.com/view/article/66096, accessed 24 May 2006.
____ (2004b) 'Motion and Time (Review)', *Viewfinder,* 57, 8–9.
____ (2005) 'Kearton, Cherry', in R. Abel (ed.) *Encyclopedia of Early Cinema.* London: Routledge, 355.
____ (2006) 'Charles Urban, Motion Picture Pioneer: Science, Education and Discovery in the Early Years of Cinema'. Available at http://www.charlesurban.com/intro.htm
Mee, A. (1911) *Harmsworth Popular Science.* London: Amalgamated Press.
Mercer, H. (1991) 'The Labour Governments of 1945–51 and Private Industry', in N. Tiratsoo (ed.) *The Atlee Years.* London: Pinter Publishers, 71–89.
M'Gonigle, G. and J. Kirby (1936) *Poverty and Public Health.* London: Victor Gollancz.
Miall, L. (1994) *Inside the BBC: British Broadcasting Characters.* London: Weidenfeld and Nicolson.
Miller, S. (2000) 'Public Understanding of Science at the Crossroads', *Science Communication, Education and the History of Science organised by the British Society for the History of Science.* Royal Society, London, 12–13 July.
Misa, T. J., P. Brey and A. Feenberg (2003) *Modernity and Technology.* Cam-

bridge, Mass: MIT Press.
Mitman, G. (1993) 'Cinematic Nature: Hollywood Technology, Popular Culture, and the American Museum of Natural History', *Isis*, 84, 637–61.
____ (1999) *Reel Nature: America's Romance with Wildlife on Film*. Cambridge, Mass: Harvard University Press.
Montagu, I (1933) 'Translator's Preface', in V. Pudovkin *Pudovkin on Film Technique: Five Essays and Two Addresses*. London: George Newnes, vii–xi.
Morell, J. B. (1990) 'Professionalisation', in R. Olby, G. Cantor, J. Christie and M. Hodge (eds) *Companion to the History of Modern Science*. London: Routledge, 980–9.
Morus, I. R. (2006) 'Seeing and Believing Science', *Isis*, 97, 101–10.
Mottram, V. H. (1949) 'Purpose in the Broadcasting of Science', *BBC Quarterly*, 3, 223–9.
Murphie, A. and J. Potts (2003) *Culture and Technology*. Basingstoke: Macmillan.
Murrell, R. K. (1987) 'Telling it Like it Isn't: Representations of Science in Tomorrow's World', *Theory, Culture and Society*, 4, 89–106.
Musser, C. (1990) *The Emergence of Cinema: The American Screen to 1907*. New York: Charles Scribner's Sons.
Neurath, M. and R. S. Cohen (eds) (1973) *Otto Neurath: Empiricism and Sociology*. Dordrecht: D Reidel Publishing Company.
Neurath, M. (1939) *Modern Man in the Making*. London: Secker and Warburg.
____ (1996) 'Visual Education: Humanism versus Popularisation', in E. Nemeth and F. Stadler (eds) *Encyclopaedia and Utopia*. Dordrecht: Kluwer, 245–335.
Newstead, M. J. (ed.) (1994) *Cumulative Index to the Quekett Journals of Microscopy*. Northolt: Quekett Microscopical Club.
Nichols, B. (1976) 'Documentary Theory and Practice', *Screen*, 17, 34–48.
____ (1991) *Representing Reality: Issues and Concepts in Documentary*. Bloomington: Indiana University Press.
Nichols, T. (1997) 'Small Screen Science: The Development of Science Programming Policy at BBC Television (1946–70)', unpublished MSc thesis, University of Manchester.
Nicholson, M. (1981a) 'The Proposal for a National Plan', in J. Pinder (ed.) *Fifty Years of Political and Economic Planning: Looking Forward 1931–1981*. London: Heinemann, 5–8.
____ (1981b) 'PEP Through the 1930s: Growth, Thinking, Performance', in J. Pinder (ed.) *Fifty Years of Political and Economic Planning: Looking Forward 1931–1981*. London: Heinemann, 32–53.
Nikolow, S. (2004) 'Planning, Democratization and Popularization with ISO-

TYPE, ca. 1945', in F. Stadler (ed.) *Induction and Deduction in the Sciences.* Dordrecht: Kluwer Academic Publishers, 299–329.
Noble, P. (ed.) (1947) *The British Film Yearbook, 1947–48.* London: Skelton Robinson.
Nyhart, L. K. (1996) 'Natural History and the "New" Biology', in N. Jardine, J. Secord and E. Spary (eds) *Cultures of Natural History.* Cambridge: Cambridge University Press, 426–43.
Olby, R. (2004) 'Huxley, Sir Julian Sorell (1887–1975)', *Oxford Dictionary of National Biography.* Oxford: Oxford University Press. Available at: http://www.oxforddnb.com/view/article/31271, accessed 18 December 2006.
Olby, R., G. Cantor, J. Christie and M. Hodge (eds) (1990) *Companion to the History of Modern Science.* London: Routledge.
Orbanz, E. (1977) *Journey to a Legend and Back: the British Realistic Film.* Berlin: Volker Spiess.
Orr, J. B. (1936) *Food, Health and Income: Report on a Survey of Diet in Relation to Income.* London: Macmillan.
____ (1941) 'Paper to Science and World Order Conference', *Advancement of Science*, 2, 26–8.
____ (1942) *Fighting for What? To Billy Boy and all the Other Boys Killed in the War.* London: Macmillan.
____ (1943) 'The Role of Food in Postwar Reconstruction', *International Labour Review*, XLVII.
____ (1966) *As I Recall.* London: Maggibon and Kee Ltd.
Orr, J. B. and S. Tallents (1958) 'Elliot, Walter Elliot', *Biographical Memoirs of Fellows of the Royal Society*, 4, 73-80.
Ostherr, K. (2005) *Cinematic Prophylaxis: Globalization and Contagion in the Discourse of World Health.* Durham, NC: Duke University Press.
Oudshoorn, N. and T. Pinch (eds) (2003) *How Users Matter: The Co-Construction of Users and Technologies.* Cambridge, Mass: MIT Press.
Pawley, E. (1972) *BBC Engineering, 1922–1972.* London: BBC.
Pearse, I. H. and L. H. Crocker (1943) *The Peckham Experiment: A Study of the Living Structure of Society.* London: George Allen and Unwin.
Pearson, D. (1982) '"Speaking for the common man": Multi-Voice Commentary in *World of Plenty* and *Land of Promise*', in P. Marris (ed.) *Paul Rotha.* London: British Film Institute, 64–85.
Pedersen, S. and P. Mandler (eds) (1994) *After the Victorians: Private Conscience and Public Duty in Modern Britain.* London: Routledge.
Perkin, H. (1989) *The Rise of Professional Society.* London: Routledge.
Peto, J. and D. Loveday (eds) (1999) *Modern Britain 1929–1939.* London: Design Museum.
Petrie, D. and R. Kruger (eds) (1999) *A Paul Rotha Reader.* Exeter: University

of Exeter Press.
Pickstone, J. V. (2000) *Ways of Knowing*. Manchester: Manchester University Press.
Pimlott, B. (1977) *Labour and the Left in the 1930s*. Cambridge: Cambridge University Press.
Pinder, J. (ed.) (1981) *Fifty Years of Political and Economic Planning: Looking Forward 1931–1981*. London: Heinemann.
Pirie, N. W. (1966) 'John Burdon Sanderson Haldane', *Biographical Memoirs of Fellows of the Royal Society*, 12, 219–49.
Plantinga, C. R. (1997) *Rhetoric and Representation in Nonfiction Film*. Cambridge: Cambridge University Press.
Porter, R. (1990) 'The History of Science and the History of Society', in R. Olby, G. Cantor, J. Christie and M. Hodge (eds) *Companion to the History of Modern Science*. London: Routledge, 32–46.
Postgate, R. (1933) *Karl Marx*. London: Hamish Hamilton.
Pound, R. (1956) 'Critic on the Hearth', *The Listener*, 12 July, 66.
____ (1958) 'Critic on the Hearth', *The Listener*, 1 November, 724.
Priestley, J. B. (1934) *English Journey*. London: William Heinemann Ltd.
Pudovkin, V. (1933) *Pudovkin on Film Technique: Five Essays and Two Addresses*, trans. I. Montagu. London: George Newnes.
Quigley, H. (1925) *Electrical Power and National Progress*. London: George Allen & Unwin.
Rabinbach, A. (1990) *The Human Motor: Energy, Fatigue, and the Origins of Modernity*. New York: Basic Books.
Reisch, G. A. (1994) 'Planning Science: Otto Neurath and "The International Encyclopaedia of Unified Science"'. *BJHS*, 27, 153–76.
Reiser, S. J. (1978) *Medicine and the Reign of Technology*. Cambridge: Cambridge University Press.
Reisz, K. and G. Millar (1972) *The Technique of Film Editing*. London & New York: Focal Press.
Richards, J. (1984) *The Age of the Dream Palace*. London: Routledge.
____ (2000) 'Rethinking British Cinema', in J. Ashley and A. Higson (eds) *British Cinema Past and Present*. London: Routledge, 21–34.
Richards, R. J. (2003) 'Biology', in D. Cahan (ed.) *From Natural Philosophy to the Sciences: Writing the History of Nineteenth-Century Science*. Chicago and London: University of Chicago Press, 16–48.
Ritchie Calder, P. (1941a) *Start Planning Britain Now*. London: Kegan Paul.
____ (1941b) *Carry On London*. London: The English Universities Press Ltd.
____ (1941c) *The Lesson of London*. London: Secker and Warburg.
____ (1950) 'Science on the Air', *BBC Quarterly*, V, 218–24.

Robinson, D. (1957a) 'Looking for Documentary. 1: The Background to Production', *Sight and Sound*, 27, 6–11.

____ (1957b) 'Looking for Documentary. 2: The Ones that Got Away', *Sight and Sound*, 27, 70–5.

Roiser, M. (2001) 'Social psychology and social concern in 1930s Britain', in G. Bunn, A. Lovie and G. Richards (eds) *169–187*. Leicester: BPS Books, 169–87.

Rosenberg, D. M., R. A. Bodaly and P. J. Usher (1995) 'Environmental and Social Impacts of Large Scale Hydro-electric Development: Who is Listening?', *Global Environmental Change*, 5, 127–48.

Ross, D. (1950) 'The Documentary in Television', *BBC Quarterly*, 5, 19–23.

Rotha, P. (1930) *The Film Till Now*. London: Jonathan Cape.

____ (1934) 'Review: *Pudovkin on Film Technique*', *Sight and Sound*, 2, 8, 140.

____ (1935) 'The Rise of Documentary', *Design for Today*, August, 313–17.

____ (1936a) *Documentary Film*. London: Faber.

____ (1936b) 'Personal View', *Manchester Evening News*, 6 November, n.p.

____ (1939) 'Letter to the Editor', *Sight and Sound*, 8, 81.

____ (1940) 'The Film at War', *New Statesman and Nation*, 230–1.

____ (1943) 'Interpreting Science by Film', *Advancement of Science*, 2, 305–7.

____ (1944) 'Films for Science', *Discovery*, 25, 109–14.

____ (1952) *Documentary Film*. London: Faber.

____ (1955) 'Television and the Future of Documentary', *Quarterly of Film, Radio and Television*, 9, 366–73.

____ (ed.) (1956) *Television in the Making*. London: Focal Press.

____ (1958) *Rotha on the Film: A Selection of Writings about the Cinema*. London: Faber and Faber.

____ (1973) *Documentary Diary: An Informal History of the British Documentary Film, 1928–1939*. London: Secker and Warburg.

____ (1982) 'Lord Ritchie Calder', *The Times*, 6 February, 8.

Royal Society of Medicine and Scientific Films Association (eds) (1948) *Catalogue of Medical Films*. London: ASLIB.

Russell, P. (2007) *100 British Documentaries*. London: British Film Institute.

Ruxin, J. N. (1996) *Hunger, Science and Politics: FAO, WHO and Unicef Nutrition Policies, 1945–1978*. London: University of London.

Ryan, T. (1980) 'Film and Political Organisations in Britain 1929–39', in D. Macpherson (ed.) *British Cinema: Traditions of Independence*. London: British Film Institute, 51–69.

Samson, J. (1986) 'The Film Society, 1925–1939', in C. Barr (ed.) *All Our Yesterdays: 90 Years of British Cinema*. London: British Film Institute, 306–13.

Scannell, P. (1979) 'The Social Eye of Television, 1946–1955', *Media, Culture & Society*, 1, 97–106.

____ (1980) 'History and Culture'. *Media, Culture & Society*, 1, 1–3.

Scott, P. (1958) 'Watching Birds and Animals', *Radio Times*, 3 January, 7.

Scientific Film Committee of the Association of Scientific Workers (1941), *The Scientific Worker* (free study leaflet).

Science Museum (2002) A Star of Astronomy. http://www.ingenious.org.uk/Read/understandingourworld/SkyWatching/Astarofastronomy/

Searle, G. R. (1979) 'Eugenics and Politics in Britain in the 1930s', *Annals of Science*, 36, 159–70.

Secord, A. (1994) 'Science in the Pub: Artisan Botanists in Early Nineteenth-Century Lancashire', *History of Science*, 32, 269–315.

Secord, J. (1996) 'The Crisis of Nature', in N. Jardine, J. Secord and E. Spary (eds) *Cultures of Natural History*. Cambridge: Cambridge University Press, 447–59.

Shapin, S. (1990) 'Science and the Public', in R. Olby, G. Cantor, J. Christie and M. Hodge (eds) *Companion to the History of Modern Science*. London: Routledge, 990–1007.

Shaw, A. G. (1952) *The Purpose and Practice of Motion Study*. Manchester: Harlequin Press Co.

Shaw, H. (1946) 'The Dissemination of Scientific information to the Public', *Advancement of Science*, 4, 26–9.

Silverstone, R. (1985) *Framing Science: The Making of a BBC Documentary*. London: British Film Institute.

____ (1986) 'The Agonistic Narratives of Television Science', in J. Corner (ed.) *Documentary and the Mass Media*. London: Edward Arnold, 81–106.

____ (1987) 'Narrative Strategies in Television Science', in J. Curran, A. Smith and P. Wingate (eds) *Impacts and Influences: Essays on Media Power in the Twentieth Century*. London: Methuen, 291–330.

Silvey, R. (1974) *Who's Listening: The Story of BBC Audience Research*. London: Allen and Unwin.

Singer, A. (1957a) 'The Restless Sphere', *Radio Times*, 28 June, 5.

____(1957b) 'Keeping an "Eye on Research"', *Radio Times*, 15 November, 7.

____ (1958) 'Eye on Research'. *Radio Times*, 26 September, 6.

____ (1965) 'The Communication of Science', *BBC Hand Book*, 25–7.

____ (1966) 'Science Broadcasting', *BBC Lunch-Time Lectures. Series 4; no. 1–6*. London: British Broadcasting Corporation.

Skal, D. A. (1998) *Screams of Reason: Mad Science and Modern Culture*. New York: W. W. Norton.

Smith, D. (1986) 'Nutrition in Britain in the Twentieth Century', unpublished PhD thesis, Edinburgh University.

____ (1992) 'Nutrition Science and Nutrition Politics during the Second World War', unpublished paper read to the Historians and Nutritionists Seminar, Kings College, London, 8 July.

____ (2000) 'The Carnegie Survey: Background and Intended Impact', in A. Fenton (ed.) *Order and Disorder: The Health Implications of Eating and Drinking in the Nineteenth and Twentieth Centuries.* East Linton: Tuckwell Press, 64–80.

Smith, D. and M. Nicolson (1989) 'The "Glasgow School" of Paton, Findlay and Cathcart: Conservative Thought in Chemical Physiology, Nutrition and Public Health', *Social Studies of Science*, 19, 195–238.

____ (1993) 'Health and Ignorance: Past and Present', in H. Thomas, S. Scott and G. Williams (eds) *Locating Health: Sociological and Historical Explanations.* Aldershot: Avebury, 221–43.

____ (1995) 'Nutrition, Education, Ignorance and Income: A Twentieth-Century Debate', in H. Kamminga and A. Cunningham (eds) *The Science and Culture of Nutrition, 1840–1940.* Amsterdam: Rodopi, 288–318.

Smith, D. C. (1986) *H. G. Wells: Desperately Mortal.* New Haven: Yale University Press.

Smith, F. P. (1907) 'The British Spiders of the Genus Lycosa', *Journal of the Quekett Microscopical Club*, Series 2, Vol X, 9–30.

____ (1908) 'Some British Spiders Taken in 1908', *Journal of the Quekett Microscopical Club*, Series 2, Vol X, 311–34.

Smith, H. L. (ed.) (1986) *War and Social Change: British Society in the Second World War.* Manchester: Manchester University Press.

Spitta, E. J. (1923) 'History of the Quekett Microscopical Society; Resumé at 500th Ordinary Meeting', *Journal of the Quekett Microscopical Society*, XV, 419–21.

Stanford, B. (1947) 'British Medical Films: A Study of the Evolution of the Medical Film in this Country', *Documentary News Letter*, 70, 79–80.

Stevenson, J. (1986) 'Planner's Moon? The Second World War and the Planning Movement', in H. L. Smith (ed.) *War and Social Change.* Manchester: Manchester University Press, 58–77.

Stollery, M. (2000) *Alternative Empires: European Modernist Cinemas and the Cultures of Imperialism.* Exeter: Exeter University Press.

Stone, L. (1971) 'Prosopography', *Daedalus*, 100, 46–79.

Sussex, E. (1975) *The Rise and Fall of British Documentary.* Berkeley: University of California Press.

Swallow, N. (1954) ' The Factual Documentary' *The Writer*, 10–11

____ (1956) 'Documentary TV Journalism', in P. Rotha (ed.) *Television in the Making.* London: Focal Press, 49–55.

____ (1966) *Factual Television.* London: Focal Press.

Swann, B. and F. Aprahamian (eds) (1999) *JD Bernal: A Life in Science and Politics.* London: Verso.
Swann, P. (1989) *The British Documentary Film Movement, 1926–1946.* Cambridge: Cambridge University Press.
Swinson, A. (1955) *Writing for Television.* London: Black.
Tagg, J. (1988) *The Burden of Representation: Essays on Photographies and Histories.* Basingstoke: Macmillan Education.
Talbot, F. A. (1912) *Moving Pictures: How They are Made and Worked.* London: Heinemann.
Taylor, P. M. (1989) 'The Projection of Britain Abroad', in M. L. Dockrill and J. W. Young (eds) *British Foreign Policy, 1945–56.* Basingstoke: Macmillan, 9–30.
Taylor, R. and I. Christie (eds) (1991) *Inside the Film Factory.* London: Routledge.
Thomson, D. (2005) *Secret Histories.* Available at: http://www.sffs.org/fest05/awards/adam_curtis.html, accessed 7 March 2007.
Thorpe, F. and N. Pronay (1980) *British Official Films in the Second World War.* Oxford: Clio Press.
Tomlinson, J. (1992) 'Productivity Policy', in H. Mercer, N. Rollins and J. Tomlinson (eds) *Labour Governments and Private Industry: The Experience of 1945–51.* Edinburgh: Edinburgh University Press, 37–54.
Tosi, V. (1960) 'The Problems of Today and Tomorrow in the Popularisation of Science Films', *Scientific Film Review,* 243–7.
____ (2005) *Cinema Before Cinema: The Origins of Scientific Cinematography.* London: British Universities Film and Video Council.
Toye, R. (2000) 'Gosplanners versus Thermostatters: Whitehall Planning Debates and their Political Consequences: 1945–1951', *Contemporary British History,* 14, 81–106.
Tripp, B. M. H. (1943) 'Science and the Citizen: The Public Understanding of Science', *Nature,* 151, 382–5.
Tubbs, R. (1942) *Living in Cities.* Harmondsworth: Penguin.
Turner, F. M. (1980) 'Public Science in Britain, 1880–1919', *Isis,* 71, 589–608.
____ (1997) 'Practicing Science: An Introduction', in B. Lightman (ed.) *Victorian Science in Context.*
Twyman, M. (ed.) (1975) *Graphic Communication through Isotype.* Reading: University of Reading.
Uebel, T. (ed.) (1991) *Rediscovering the Forgotten Vienna Circle.* Dordrecht: Kluwer Academic Publishers.
Urban, C. (1906) *November 1906 Supplement to June 1905 Charles Urban Trading Company Catalogue.* Republished on CD, c. 2004, Hastings: The Projec-

tion Box.
____ (c.1906) *The Cinematograph in Operative Surgery (Charles Urban Trading Company CD Archive Volume One)*. Hastings: The Projection Box.
____ (1907) *The Cinematograph in Science, Education and Matters of State*. London: Charles Urban Trading Company.
____ (1909) *General Catalogue of Classified Subjects*. London: Charles Urban Trading Company.
Various (1942) 'Science and World Order', *Advancement of Science*, 2, 1–120.
Vassilkov, I. (1959) 'The Popular Science Film: Subject, Purpose and Means of Popularisation', *Science and Film*, 2/3, 54–64.
Vesselo, A. (1936) '*Enough to Eat?* (The Nutrition Film)', *Monthly Film Bulletin*, 3, 187–8.
Vincent, B. B. (1997) 'In the Name of Science', in J. Krige and D. Pestre (eds) *Science in the Twentieth Century*. Amsterdam: Harwood Academic Publishers, 319–38.
Waters, C. (1994) 'J. B. Priestley 1894–1984: Englishness and the Politics of Nostalgia', in S. Pedersen and P. Mandler (eds) *After the Victorians: Private Conscience and Public Duty in Modern Britain. Essays in Memory of John Clive*. London: Routledge, 209–26.
Waters, C. K. and A. Van Helden (1992) *Julian Huxley: Biologist and Statesman of Science*. Houston: Rice University Press.
Watson Watt, R. A. and B. A. Keen (1936) 'Films and Science. A Symposium', *Sight and Sound*, 5, 150–1.
Weber, M. (1948) *From Max Weber: Essays in Sociology*. London: Routledge & Kegan Paul.
Webster, C. (1982) 'Healthy or Hungry Thirties?', *History Workshop Journal*, 13, 110–29.
Wells, H. G. (1939) *Peace and War Aims. A Correspondence from The Times. (Including H. G. Wells' New Declaration of "The Rights of Man")*. London: Peace Book Co.
Wengenroth, U. (2003) 'Science, Technology and Industry', in D. Cahan (ed.) *From Natural Philosophy to the Sciences: Writing the History of Nineteenth-Century Science*. Chicago and London: University of Chicago Press, 221–53.
Werskey, G. (1971) 'British Scientists and "Outsider" Politics, 1931–1945', *Science Studies*, 1, 67–83.
____ (1988) *The Visible College*. London: Free Association Books.
Whitby, J. (1996) 'The Future of the Cinematograph', in C. Harding and S. Popple (eds) *In the Kingdom of Shadows: A Companion to Early Cinema*. London: Cygnus Arts, 21–2.

White, H. V. (1974) *Metahistory. The Historical Imagination in Nineteenth-Century Europe.* Baltimore & London: Johns Hopkins University Press.
Wildy, T. (1986) 'From MOI to COI: Publicity and Propaganda in Britain 1945–51: The National Health and Insurance Campaigns of 1948', *Historical Journal of Film, Radio and Television*, 6, 3–19.
____ (1988) 'British Television and Official Film, 1946–1951', *Historical Journal of Film, Radio and Television*, 8, 195–202.
Williams, R. (1976) *Keywords: A Vocabulary of Culture and Society.* London: Fontana.
____ (1992) 'When Was Modernism?', in F. Frascina and J. Harris (eds) *Art in Modern Culture.* London: Phaidon Press, 23–7.
Williams, T. I. (2004) 'Calder, Peter Ritchie, Baron Ritchie Calder (1906–1982)', *Oxford Dictionary of National Biography.* Oxford: Oxford University Press. Available at: http://www.oxforddnb.com/view/article/30891, accessed 12 September 2007.
Winnington, R. (1975) *Film Criticism and Caricatures 1943–53*, ed. P. Rotha. London: Paul Elek.
Winser, J. K. (1940) 'Review of Science in War', *Architects' Journal*, 92, 134–5
Winston, B. (1995) *Claiming the Real: The Documentary Film Revisited.* London: British Film Institute.
____ (2000) *Lies, Damn Lies and Documentaries.* London: British Film Institute.
Wood, N. (1959) *Communism and British Intellectuals.* London: Victor Gollancz.
Woolfe, B. (1941) 'I remember', *Sight and Sound*, 10, 8–9.
Young, M. (1981) 'The Second World War', in J. Pinder (ed.) *Fifty Years of Political and Economic Planning: Looking Forward 1931–1981.* London: Heinemann, 81–96.
Young, R. M. (1986) 'The Dense Medium: Television as Technology', *Political Papers*, 3–5. Available at: http://www.human-nature.com/rmyoung/papers/paper96.html, accessed 6 February 2007.
____ (1995) 'What I Learned at Summer Camp: Experiences in Television.' Talk given at the Department of the History and Philosophy of Science, University of Cambridge, 1995. Available at: http://human-nature.com/rmyoung/papers/paper29.html, accessed 6 February 2007.
Zguridi, A. and B. Altschuler (1959) 'The Classification of Scientific Films', *Science and Film*, 8, 51–4.
Ziegler, P. (2004) 'Zuckerman, Solly, Baron Zuckerman (1904–1993), *Oxford Dictionary of National Biography.* Oxford: Oxford University Press. Available at: http://www.oxforddnb.com/view/article/53466, accessed 14 Sep-

tember 2007.
Ziman, J. (1991) 'Public Understanding of Science', *Science, Technology & Human Values*, 16, 99–105.
Zuckerman, S (as Anon.). (1940) *Science in War*. Harmondsworth: Penguin.
____ (1978) *From Apes to Warlords: an Autobiography, 1904–46*. London: Collins.

INDEX

Nonfictions is dedicated to expanding and deepening
the range of contemporary documentary studies.
It aims to engage in the theoretical conversation
about documentaries, open new areas of scholarship,
and recover lost or marginalised histories.

General Editor, Professor Brian Winston

Other titles in the *Nonfictions* series:

Direct Cinema: Observational Documentary and the Politics of the Sixties
by Dave Saunders

Projecting Migration: Transcultural Documentary Practice
edited by Alan Grossman and Àine O'Brien

The Image and the Witness: Trauma, Memory and Visual Culture
Edited by Frances Guerin and Roger Hallas

Vision On: Film, Television and the Arts in Britain
by John Wyver

Building Bridges: The Cinema of Jean Rouch
edited by Joram ten Brink

Forthcoming titles in the *Nonfictions* series:

Documentary Display: Re-Viewing Nonfiction Film and Video
by Keith Beattie

Chávez: The Revolution Will Not Be Televised – A Case Study in Politics and the Media
by Rod Stoneman